TETRA'S *Popular* GUIDE TO TROPICAL CICHLIDS

TETRA's *Popular* GUIDE TO TROPICAL CICHLIDS

Tetra Press

A Salamander Book

Published in North America by Tetra Press,
3001 Commerce Street
Blacksburg VA 24060

This edition © 1994 Salamander Books Ltd

ISBN 1-56465-147-9

This book may not be sold outside the United States and Canada

All rights reserved. No part of this book may be reproduced, stored in a retrieval system or transmitted in any form or by any means, electronic, mechanical, photocopying, recording or otherwise without the prior permission of Salamander Books Ltd.

All correspondence concerning the content of this volume should be addressed to Salamander Books Ltd, 129–137 York Way, London N7 9LG, United Kingdom.

Credits

Editors: Tony Hall, Coral Walker

Designer: Paul Johnson

Color artwork: Bill Le Fever, Stephen Gardner, Brian Watson (Linden Artists), Clifford and Wendy Meadway, Colin Newman (Linden Artists), Gordon Munro. © Salamander Books Ltd

Color reproductions: P&W Graphics PTE Ltd, Singapore; Bantam Litho Ltd; Melbourne Graphics Ltd.

Filmset: SX Composing Ltd

Printed in Belgium by Proost International Book Production

Authors

David Sands has studied fishes for almost thirty years and is a regular contributor to aquarist magazines in the UK and USA. He is currently completing research on fish behavior for a PhD at Liverpool University. As an importer and retailer of tropical freshwater and marine fishes, David has encountered at first hand all the challenges which face the fishkeeper. He has made frequent field trips to study and photograph tropical fishes in their natural habitats.

Dr Paul V Loiselle's interest in aquarium fishes dates back over twenty years. He took his Master's Degree at Occidental College in Los Angeles and obtained his doctorate from the University of California at Berkeley. During five years as a Peace Corps fisheries biologist in West Africa, he carried out faunal and environmental surveys in Togo and Ghana and has since returned to Africa to make first-hand observations of cichlid behaviour in Lakes Victoria and Tanganyika. He is a founder member and Fellow of the American Cichlid Association and Technical Editor of the ACA's Journal.

Dr Wayne S Leibel has been an aquarist for over 35 years, specializing in fish of South and Central America for the past 15 years. He took his Master's and Doctorate Degree from Yale University in comparative biochemistry and was a Research Fellow at Harvard University before taking up his present position as an Associate Professor of biology at Lafayette College in Pennsylvania USA. Dr Leibel has served two terms on the Board of the American Cichlid Association and for seven years edited their Journal *The Buntbarsche Bulletin.* He was recently made a Fellow of the ACA.

Contents

The *Fishkeeper's Guides* cichlid series, a success on both sides of the Atlantic, proved the real popularity of this group of fish with aquarists. Cichlids and catfish currently dominate experienced aquarist's aquaria just as they do in the freshwaters of Africa, South and Central America. Both major groups of fish have certainly dominated the ornamental fish industry and hobby for the last three decades.

In this new publication, as in the original best-selling books, Dr Wayne Leibel, Dr Paul Loiselle and myself provide newcomers and established aquarists with a wealth of information on the three major cichlid groups, both factual and experience related. This combined expertise spans so many years and is enhanced by each expert's fishkeeping knowledge. This blend of information, from three established authors and researchers is brought together into a single volume for the first time. The results of this publication can only continue to assist fishkeepers in understanding the importance of maintenance and husbandry of all popular cichlids.

Aquarists involved in the day-to-day care of fish willingly and sometimes unwittingly encounter the wonderful world of cichlid behaviour, especially in reproductive and parental matters. Most fishkeepers are called upon to embrace general subjects such as water chemistry, geology, geography, nonmenclature and taxonomy. This work takes these factors and reveals the understanding of them in the best manner possible. Enquiring aquarists want to know as much as possible about the cichlids in their care and I suspect that this book will become a real favourite amongst those aquarists with a yearning to know more than just the basics. The most popular cichlid species are pictorially detailed within each particular section. They are dealt with in a concise style, by each author, in an easy to access sequence of text.

African cichlids, especially the Lakes Malawi and Tanganyika species, have enjoyed fluctuating popularity over the years although there have always been serious aquarists who have found the Lake cichlids irresistible. Some of the riverine species have a

fixed place in fishkeeping in that very few community aquaria have not been home to *Pelvicachromis*, better known as 'Kribs'. Dr Loiselle deftly examines the three main groups of African cichlids in the way only he could and the hobby can only be richer for his style.

Dr Leibel carefully brings home the incredible range of South American cichlids, from Oscars to Chequerboards, examining the 'King of the Amazon', better known as Discus on one side and the 'Predators', better known as Pike Cichlids, on the other. The discus aquarium is widely accepted as the ultimate challenge for aquarists and Dr Leibel provides important comments for the cichlid enthusiast wanting to venture into this demanding community of elegant fish.

Central American cichlids attract fishkeepers with giant aquaria who wish to bring together a community of big fish. The range of species widely available today offers yet another great challenge to

aquarists. As with my co-authors, I have tried to include practical advice which may be lacking in the large scale pictorial books available today.

Three individualistic authors will always provide differing opinions although there is one subject we can agree on. Keeping cichlids in good health is often about keeping water. If aquarists can guarantee clean water quality and an imaginative diet for their fish then cichlids will rarely disappoint them. We offer sound advice, based on an excellent store of practical and reference information, to help towards this goal. Our summary of opinions and facts can be found within these bright pages. Finally, the high standard of text in the original cichlid editions has made editing this work a pleasure. I congratulate my fellow ichthyologists on their splendid efforts.

David Sands

One of the greatest fallacies in fishkeeping is that the hobby is chore-free and that all the enthusiast has to do is sit and watch the fish create a pretty picture as they swim by. Like any captive animal in an artificial environment, all fish need a clean home and the correct diet. Keeping an aquarium 'clean' and providing water quality is a primary task for the diligent aquarist. Cichlids in captivity are very demanding, especially when they are feeding on a broad and healthy diet – the result of which is plenty of waste to pollute the aquarium water. This is where efficient filters, accurate heater/thermostats and a strict regime of partial water changes will ensure crystal waters for healthy fish. There are ways of making the care and cleaning tasks easier and, with

this in mind, the following sections have been written to help make the work easier. Always purchase the best filters and general equipment and never be without those essential accessories: gravel cleaners, siphon sets and clean buckets. Make the task easier by keeping a good range of remedies and water test kits handy and never shirk that water change when it is due. The cichlids in your keep will thank you for any care and affection. By following the expert advice offered by three experienced aquarists, all of whom have helped spawn and raise a great many fine cichlid species over the years, success is almost bound to follow. The day that pair of cichlids bring out their brood of fry will make you proud. It will be no more than your hard work deserves.

AFRICAN CICHLIDS

For practical purposes, it is convenient to recognize three geographical subsets of African cichlids. The first group consists of those species native to the rivers of Africa. Cichlids of the northern Great Lakes, such as Lake Victoria can be included in this group because they require similar general care. A quick glance at the map overleaf will reveal that riverine cichlids are found over a tremendous area, which is divided into a number of discrete regions by ichthyologists. The second major group consists of cichlids native to Lake Malawi and in the third group are those cichlids found only in Lake Tanganyika.

Riverine cichlids
It is important to distinguish between rivers that flow for most, or all, of their length through savanna (grassland) habitats, and those that drain forested regions.

Savannas are characterized by highly seasonal patterns of rainfall. The rainy season is brief; a year's worth of rain falls over the course of 3-4 months. As a result, savanna rivers vary enormously in their rate of flow and chemical make-up during the course of one year. In the rainy season they become veritable torrents that may leave their beds and inundate enormous tracts of land. At this time, pH values in the main channel of the river hover closely around 7.0 (ie neutral). Depending upon local conditions, they can drop as low as 6.0 (ie acidic) in quiet-water habitats on the river's flood plain. Due to the tremendous influx of rainwater, hardness values seldom exceed 3°dH.

During the dry season, however, conditions change dramatically.

Even larger rivers, such as the Niger and the Zambezi, may cease to flow altogether along their upper and middle reaches. What was formerly a mighty torrent is reduced to a series of pools connected by a trickle of water! Evaporation concentrates dissolved substances significantly, so total hardness values as high as 10°dH are not unusual. As a rule, pH values range between 7.0 and 7.8 in unplanted habitats but can soar as high as 9.0 (ie alkaline) during the day in heavily vegetated oxbows (the water within a bend in

Below: *The Napoleon Gulf of Lake Victoria. Cichlids from this region appreciate neutral to slightly alkaline water and tolerate a range of hardness values.*

Basic water chemistry: pH and hardness

The pH of water

The degree of acidity or alkalinity of water is expressed in terms of pH value, which literally means 'hydrogen power'. The scale is based inversely on the concentration of hydrogen ions in the water; the more hydrogen ions, the more acid the water and lower the pH value. The pH scale ranges from 0 (extremely acidic) to 14 (extremely alkaline), with a pH value of 7 as the neutral point.

pH scale

0	1	2	3	4	5	6	7	8	9	10	11	12	13	14
extremely acidic							neutral							extremely alkaline

The scale is logarithmic, which means that a pH value of 8 represents a ten-fold decrease in the hydrogen ion concentration compared to a pH value of 7. An apparently small change in pH, from say 6 to 8, therefore represents a hundred-fold decrease in the hydrogen ion concentration, which can cause severe stress to many fish. A number of different tests for measuring pH are available. These include paper strip indicators, liquid pH test kits and more sophisticated electronic pH meters.

Water hardness

Water hardness is related to the amounts of dissolved salts present in the water. Two types of hardness are important to the fishkeeper; *total* or *general hardness* (or GH), which is related to the levels of calcium, and magnesium in the water, and *carbonate hardness* (or KH), which is related to the amounts of carbonate/ bicarbonate present. Water hardness is measured by several different scales, including degrees of German hardness (dH°). On this scale, water with a hardness value of 3°dH or less is termed 'soft' (i.e. low in dissolved salts) and water with a hardness value of over 25° dH is termed 'very hard' (i.e. rich in dissolved salts). An alternative scale is based on milligrams of calcium carbonate ($CaCO_3$) per litre, also expressed as parts per million (ppm). Again, test kits are available from aquarium dealers.

Water hardness in comparative terms

dH°	Mg/litre $CaCO_3$	Considered as:
3	0-50	Soft
3-6	50-100	Moderately soft
6-12	100-200	Slightly hard
12-18	200-300	Moderately hard
18-25	300-450	Hard
Over 25	Over 450	Very hard

the river), or in a peripheral swamp.

Fishes that can prosper in such a setting must possess considerable resilience in the face of fluctuating pH and hardness values. They are not likely to be very demanding with regard to the chemical make-up of their water in captivity, as long as extremes of pH or hardness are avoided. Not surprisingly, cichlids native to these habitats, such as *Tilapia zillii, Oreochromis mossambicus, Chromidotilapia guntheri* or *Haplochromis callipterus* have a well-established reputation for hardiness among enthusiastic aquarists!

In forested regions the situation is quite different. A tropical rain forest occurs only where there is more or less constant rainfall. The 'dry' season in such areas simply refers to the time of the year when less rain falls, contrary to the case on the savanna. Forest rivers thus enjoy a much steadier flow and are not characterized by dramatic seasonal variation in pH or hardness values.

Their waters are characterized by very low total and carbonate hardness; values in excess of 2°dH would be considered remarkable under most circumstances. When very soft waters come into contact with significant deposits of decomposing plant matter, they pick up large quantities of organic acids. Such 'blackwater' streams – so-called because of the characteristic coloration of their waters – can have pH values as low as 4.0. Even in so-called 'clearwater' streams, pH values between 5.5 and 6.2 are commonly encountered, while those in excess of 7.0 are virtually unheard of.

Most cichlids native to forest streams do not seem to object to moderately hard, neutral to slightly alkaline water conditions in captivity.

In Lake Victoria and other northern Great Lakes, water conditions resemble those characteristic of large, savanna-draining rivers during the rainy season. Lake Kivu, however, is characterized by hard alkaline water conditions, comparable to those of the two best known of the African Rift Lakes, Tanganyika and Malawi. Nevertheless, cichlids from Lakes Victoria and Kivu will thrive in neutral to slightly alkaline water over a considerable range of hardness values. Avoid abrupt changes in the chemical composition of their water.

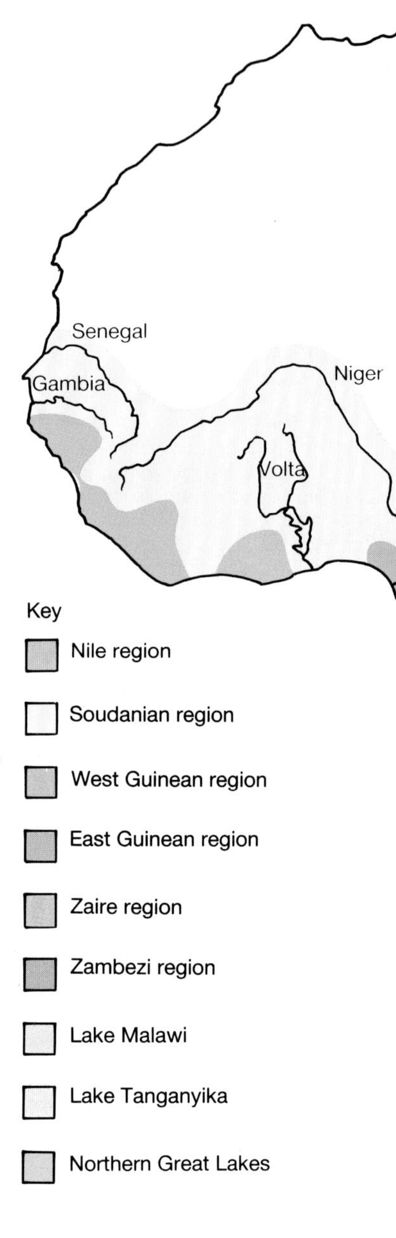

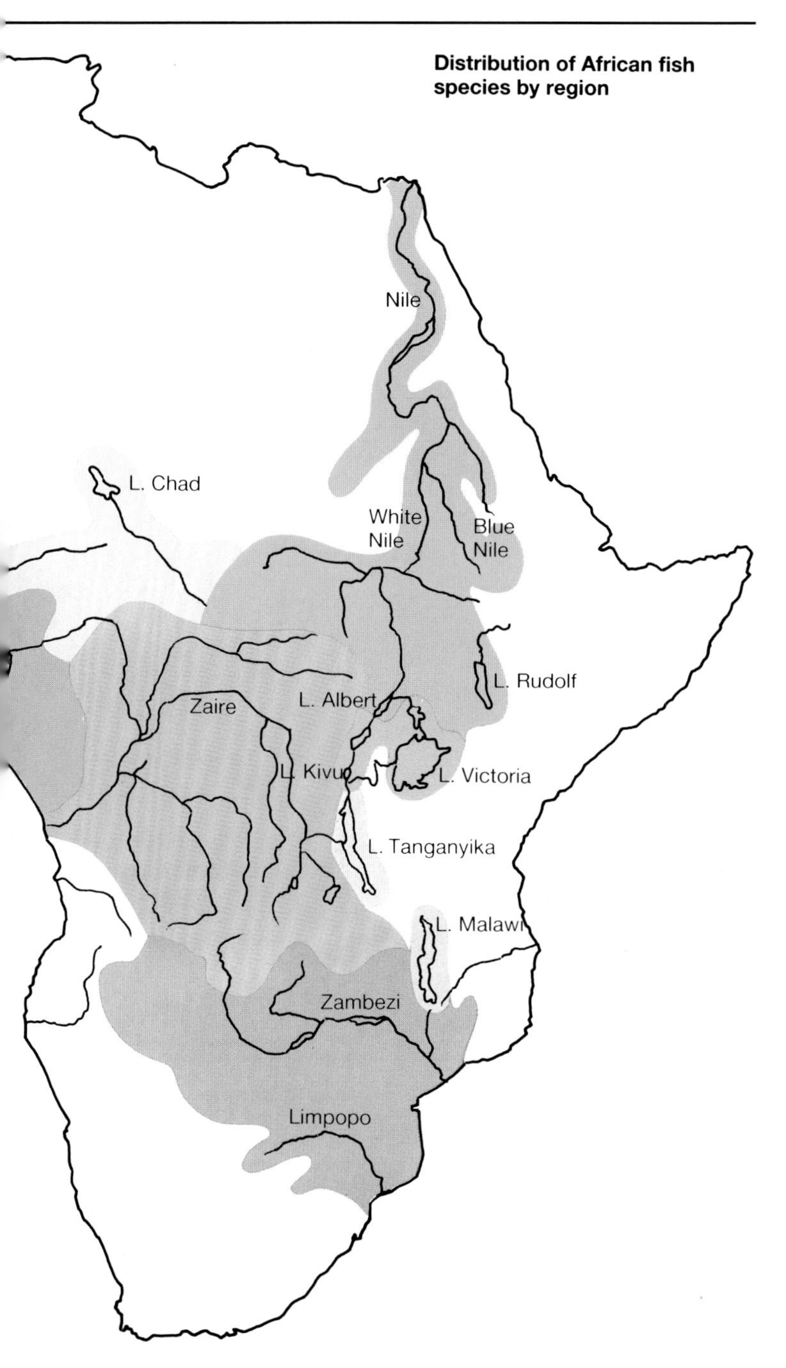

Lakes Malawi and Tanganyika
Cichlids found in these two lakes form the second and third groups, respectively. They are distinctive in that their catchment areas have been geologically active in the relatively recent past. Their rocks are much younger than those

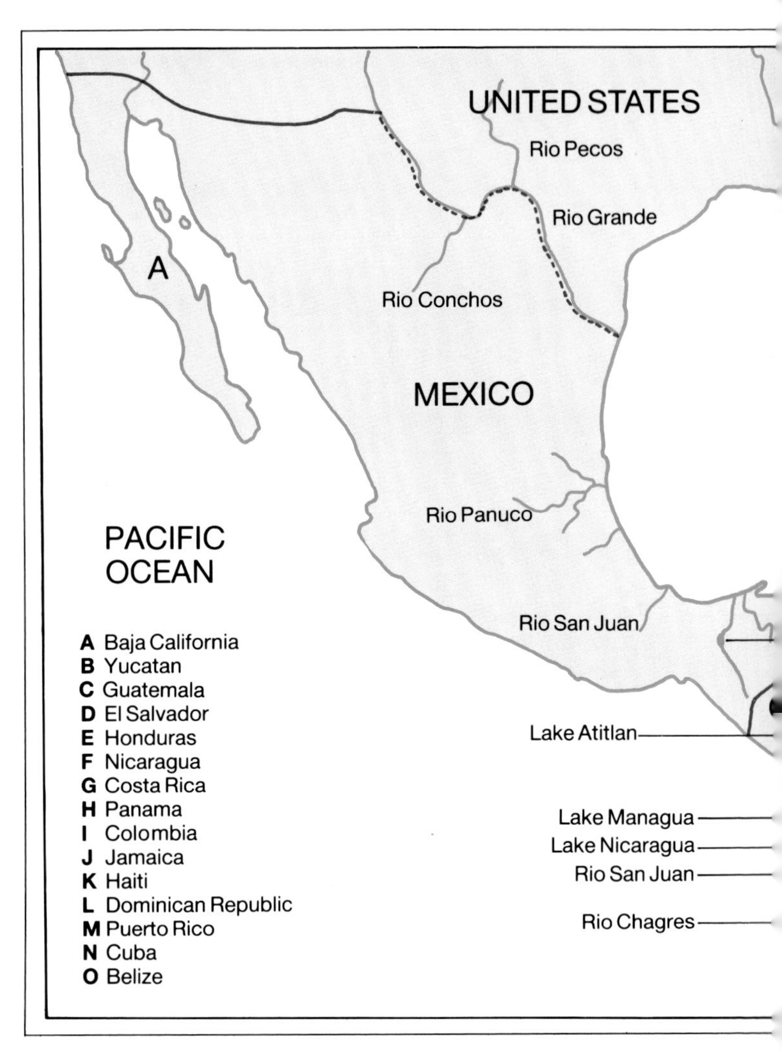

found in most of western Africa and correspondingly richer in soluble minerals. Additionally, for most of their geological histories, both lake basins have been closed, ie with no outlets to the sea.

Recorded pH values ranges from 7.7 to 8.6 in Lake Malawi, and from 7.3 to 8.0 in Lake Tanganyika. Values of 10-12°dH total hardness are recorded for Lake Tanganyika. The waters of Lake Malawi are somewhat less mineralized, with overall hardness values ranging from 6-10°dH.

Both lakes are remarkably stable in terms of chemical composition and water temperature. Cichlids native to these inland seas cope poorly with abrupt environmental changes. Tanganyikan cichlids are particularly sensitive to this.

Malawi cichlids prosper in water considerably harder and more alkaline than that of their native lake, while Tanganyikan cichlids will live and breed under softer, less alkaline conditions. Their only absolute requirements are that the pH in their tanks be alkaline and that they be gradually acclimatized to the conditions.

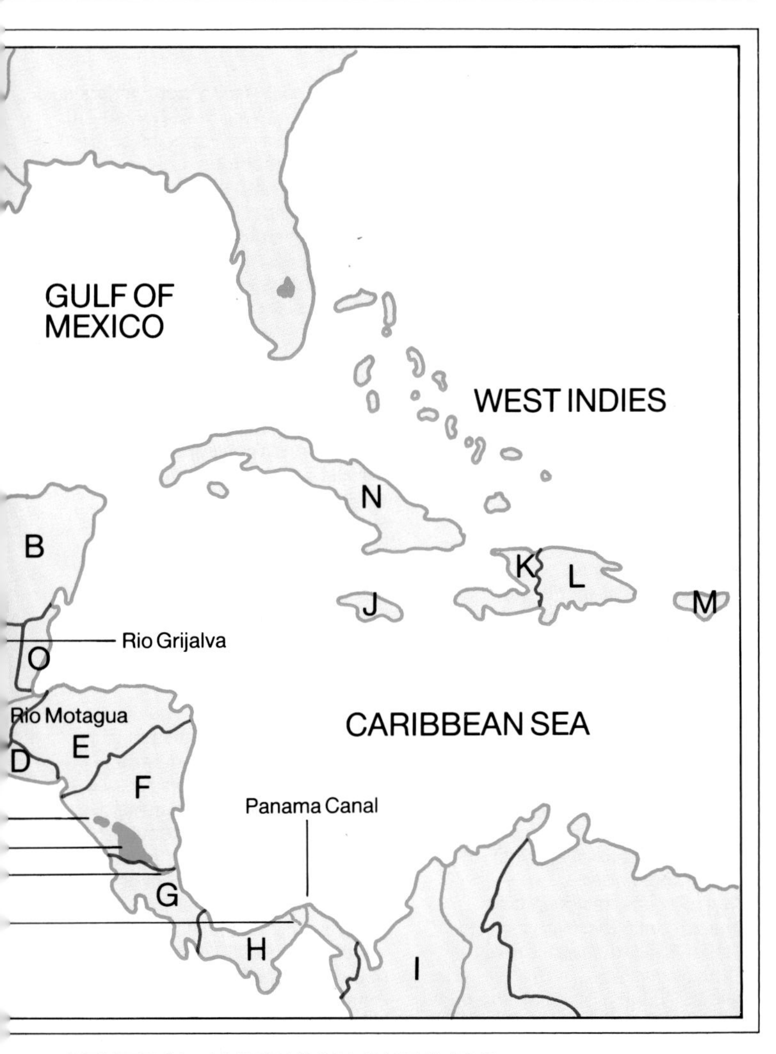

CENTRAL AMERICAN CICHLIDS

The narrow land bridge of Central America stretches south-east from Mexico through Guatemala, Belize, Honduras, El Salvador, Nicaragua, Costa Rica, Panama to Colombia in South America. It covers approximately 520,000 km (200,000 square miles) and is veined with rivers and marked with small lakes. These waterways provide a very wide range of environments; some small tributaries of major rivers are fast flowing, others are idle creeks. The flowing water, unless in full torrent, is usually clear, often agitated as it passes over rock, rubble and small boulders. The lakes in this region vary in depth, sometimes according to season and the amount of run-off from the many rivers.

The rivers and crater lakes of the Central American region are generally hard and alkaline, although some small creeks in the area are acidic and can become more so in the dry season.

SOUTH AMERICAN CICHLIDS

South America as we know it today is half the size of Africa and one-third the size of Eurasia, being some 7250km (4500 miles) long and 4800km (3000 miles) across at its widest point – a total area of approximately 18 million square km (7 million square miles). Its geography is dominated by the world's largest rainforest which is drained by the second longest river in the world, the Amazon. The Amazon River delivers one-fifth of the total riverine freshwater discharge of this planet, the largest volume of water discharge and 7-10 times that of the Mississippi River. It drains two-fifths of South America, including Peru, Ecuador, Colombia, Venezuela, the Guianas, and Brazil. It boasts over 1000 tributaries, 17 of them over 1600km (1000 miles) long. In places its depth reaches 90m (295 feet), allowing ocean-going ships access to the Peruvian port of Iquitos, some 3700km (2300 miles) inland at the foot of the Andes, and it is as wide as 11km (7 miles) at some points.

Goulding (1990) estimates that a total 2500-3000 species of fish, only half of which have yet been formally described by ichthyologists, reside in the Amazon drainage. With a world total of 6650 freshwater fish, the Amazon basin is the world's riches ichthyofaunal region with nearly 10 times as many fish species as all of Europe. Of that total, Lowe-McConnel (1991) suggests that nearly 225 species, 75 per cent of the total estimated 300 species of South American cichlids, are found in the Amazon drainage. However, cichlids make up only about 6-10 per cent of the total fish diversity of the Amazon, with characoids (tetras, silver dollars, etc. 43 per cent) and catfishes (Siluriformes: 39 per cent) being the most highly-radiated and represented groups.

Most South American cichlids are lentic fishes, inhabitants of slow-moving or stagnant waters, and, with certain expectations, they tend not to frequent the rapidly-flowing channels of rivers. Rather, they are found in small water bodies – tributaries, streams, creeks, or associated pools, oxbow lakes, laguna, marshes and the like. They get there during the rainy season, which lasts from November to June, when the Amazon rises as much as 15m (50ft) and spills over its banks to flood adjacent forest. The flooding extends as much as 80-95km (50-60 miles) beyond the normal river channel and encompasses an area of some 100 square km (38,600 square miles) – about 2 percent of the total Amazon rainforest (*igapo*) and an area larger than England. This flooding happens because the

Below: *During highwater, the rivers spill over their banks to flood the forests.*

Amazon basin is so flat, rising no more than 200m (650ft) above sea level at its highest point. Thus, the massive volume of water caused by the torrential rains and melting ice in the Andes Mountains flows eastward towards the Atlantic Ocean and spills over the river banks that are unable to contain the swell. This seasonal flooding, more than anything else, dictates the rhythm of life in Amazonia. The flooded forests are extremely important spawning grounds for most Amazonian fish, including cichlids, providing food in the form of insects and fallen fruits, and cover for newly-hatched fry. After the peak of the flooding, some fish simply escape the draining forest by returning to the river channel. Others remain in marshes or *varzea* lakes during the subsequent dry season. These seasonal floods cover vast expanses of Amazonia for 4-7 months each year.

Although it is central, the Amazon Basin is not the entire story of South America and its fishes. The Amazon Basin is bounded by the Andes Mountains to the west, the Brazilian Highlands to the south and east, and the Guiana Highlands to the north. To the west of the Andes, great deserts, 2570km (1600 miles) of them, extend from the mountains down to the sea. The Brazilian Highlands are formed of ancient crystalline rock and cover 1,500,000 square km (580,00 square miles) of primarily and savannah consisting of scrub forest and grasslands, of which the Mato Grosso is the most famous. As the plateaux diminish in altitude towards the south, the tropical deciduous forest and scrub of the Gran Chaco dominate the landscapes of Bolivia, Paraguay and northern Argentina. The partly swampy plains or *chaco* grade southward into the Argentine *pampas*, or vast arid grassland regions. To the south, along the eastern base of the Argentine Andes, is a large desert with shrubs and cacti. To the south and

Above: *During the rainy season, the rivers rise as much as 15 metres (50 feet). The mud coating on this tree denotes the highwater mark.*

west, in the shadow of the Chilean Andes, humid temperate (subtropical) forests characterise the region known as Patagonia which, as you travel southward, increasingly resembles Alaska, complete with glaciers. Southernmost South America, Tierra del Fuego, is decidedly cold and humid, and is the winter home to several species of penguins.

Clearly South America is a continent of extreme contrasts: of tropical rainforests which occupy nearly half of the landmass, but also of mountains and deserts, of huge grassland savannahs and subtropical deciduous forests. While the average yearly temperature of Amazonia is 27°C (81°F) (with extremes of 20-37°C (69.4-98.4°F) recorded at Manaus, Brazil), Uruguay to the south is decidedly temperate, averaging 10°C (50°F) in winter and 22°C (71°F) in summer. Average yearly temperature variations from 0-9°C (32-49°F) have been reported from Tierra del Fuego at the tip of the continent. Not surprisingly, South America consists of a range of 'life zones' supporting a wide variety of organisms adapted to the highly variable topography and associated climate. Since rivers penetrate most of the South American continent, it should come as no surprise that the fishes

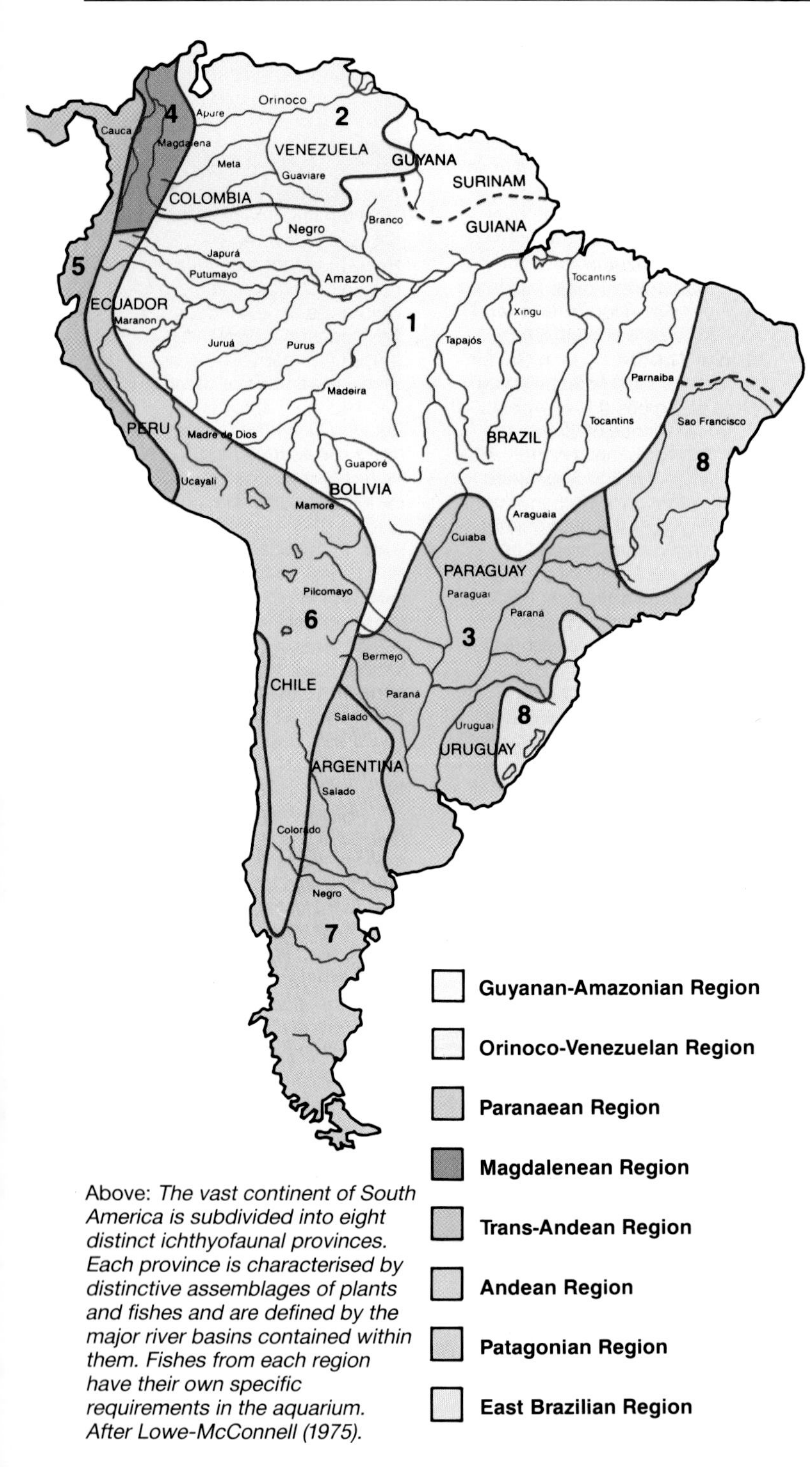

Above: *The vast continent of South America is subdivided into eight distinct ichthyofaunal provinces. Each province is characterised by distinctive assemblages of plants and fishes and are defined by the major river basins contained within them. Fishes from each region have their own specific requirements in the aquarium. After Lowe-McConnell (1975).*

of South America are equally diverse and adapted to these zones.

The French ichthyologist Jacques Gery recognized eight major ichthyofaunal regions in South America.

Guyanan-Amazonia and Orinoco-Venezuela

The Amazon drainage is more correctly known as the Guyanan-Amazonian region because the Guianas (Guyana, French Guiana, Surinam) and the Amazon basin share fish across the seasonally-flooded savannas of Guyana to the north. Similarly, the second major region, the Orinoco-Venezuelan region to the northwest, is also connected to the Amazon system via the Casiquiare Canal (Rio Negro, Brazil; Rio Orinoco, Venezuela) during highwater. The Rio Orinoco, which originates in Colombia and travels north through Venezuela to the Caribbean Sea, flows through the flat, often marshy grasslands, or *llanos*, which flood during the rainy season and exchange fish with the Rio Negro. Nearly 90 per cent of the cichlid fauna is distributed between the two regions.

Paranean region

The third major ichthyofaunal region is the Paranean. Cichlids are also found here, though not with the richness found in Amazonia. It is the second largest drainage in South America and comprises the La Plata-Rio Uruguai-Rio Parana-Rio Paraguai system in Paraguay, Uruguay and Argentina. In fact, the confluence of these rivers at La Plata makes this compound 'river' (Rio de la Plata) the second largest in the world, after the Amazon. The Paranean region is very dry and seasonally swampy, and vast marshes known as the Pantanal flood annually. Although isolated today, it was connected with the Amazon, so many of the fish species are now widely distributed from the Rio Orinoco down to this essentially temperate zone.

East Brazilian region

To the north and east of the Paranean region, and forming the eastern boundary of Amazonia (the Brazilian Highlands), is the East Brazilian ichthyofaunal region characterized by smaller rivers. While the immediate coast is humid and covered by dense forest, the highlands themselves are arid plateaux which are largely scrub forest or even desert. Many of the coastally-distributed cichlids venture into partially-saltwater estuarine habitats and can tolerate, even require, harder waters.

Other regions

The four remaining regions are not particularly rich in fish – or cichlid – species. The Magdalenean and Trans-Andean regions of northwestern South America, Colombia and Ecuador are defined by the Andes. The Magdalenean is east of these mountains and the Trans-Andean is west. There are some, but not many, cichlids found here. The Andean and Patagonian regions are nearly completely devoid of fishes due to the exceedingly low temperatures.

Cichlids water chemistry parameters (sometimes extreme) obtained from habitat research in nature should only be used as guidelines for aquarium care. Discus (an ideal example), living in an *open water* system in a tributary of the Rio Negro that has a pH of 4.7 and zero hardness would not necessarily thrive if the same conditions were established in an aquarium. In such an artificial 'closed' system, the low pH would result in low oxygen levels and the lack of any buffering capacity could result in a pH collapse, both of which would lead to health problems. Fishes are stressed by the extra pressure placed on their gills to extract oxygen from the extreme aquarium water chemistry.

Many cichlids (especially Discus) are commercially raised for the aquarium market and are not maintained in conditions parallel to water chemistry found in nature.

Aquarium Selection

General fishkeeping practice usually dictates that individual fishes are introduced into the same system over a period of time. The result is that a dominant individual often rules the community, and in certain situations, will relentlessly bully weaker, smaller cichlids. The blame for this brutality is usually placed on the fishes, but this illogical thinking hides the fact that the responsibility really lies with the fishkeeper. It is imperative for the fishkeeper to choose the species to be mixed and the numbers that can be reasonably accommodated within the optimum aquarium size. The first parameter is dictated by the species *available* in local or nearby aquarium shops; the second by the space and funds at hand. If you are considering keeping the territorial minded cichlids, these basic points are crucial.

Regardless of size, an aquarium for cichlids also requires a tight fitting cover. Cichlids, unlike some other fishes, do not jump in pursuit of food, or with the object of migrating towards some predetermined destination. However, a cichlid on the losing end of a serious quarrel may well resort to a frantic leap up and out of its tank. This tactic may allow a defeated fish to escape further harm in nature, but in captivity few cichlids benefit from a prolonged sojourn on a rug or concrete floor. The knowledge that such a loss is so easily prevented merely serves to heighten the distress of an aquarist.

Another point to consider is that most cichlids live in fairly close association with the bottom, therefore it pays to select relatively shallow tanks with an extensive base area, rather than taller, so-called 'show tanks'. An exception to this rule is the tank for the mbuna of Lake Malawi; under aquarium conditions they will happily swim among vertical rock faces. A number of pelagic species endemic to Lake Tanganyika, such as the several *Cyprichromis* species and the featherfins of the genera *Cyathopharynx* and *Ophthalmotilapia*, also appear to appreciate deeper tanks. For the most part, however, cichlids make little use of the upper two-thirds of the water column in their tank.

Cichlids produce copious quantities of nitrogenous wastes, so the tanks that house them are more susceptible than most aquaria to explosive algal growth. Many African cichlids have marked herbivorous tendencies, so relying on aquarium plants to soak up this abundance of fertilizer is not always possible. Try to locate the tank to control the lighting. It is much easier to manage algal growth if the tank is placed away from any source of natural light. Failing this, north – or east – facing windows are acceptable locations in the northern hemisphere. Tanks facing south or west are a standing invitation to unsightly algal blooms.

Finally, position a tank as close as possible to an electrical outlet. Avoid trailing extension leads, both for aesthetic and safety reasons. At one time, it was important to site the aquarium close to a drain to simplify routine maintenance. Today, the availability of automatic water changing and/or gravel cleaning devices makes it no longer essential, although undeniably convenient, to locate the tank near a water supply.

Tank size

In an ideal world, the selection of the perfect aquarium for Central American cichlids would be determined by the largest space available in the home. For these

inhabitants of the rivers, lakes, swamps, lagoons and crater lakes need plenty of space in which to exercise various postures, such as territorial defence, sexual display and dominance behaviour. In the waters of the Rio Grande, or Rio Negro tributaries juvenile cichlids will shoal in great numbers. A spawning pair of large Central American cichlids will defend a pit up to 3m (10ft) across from all-comers that threaten them.

Many cichlids, especially the African and Central American species, require large tanks if they are to prosper. In fact, giant predatory cichlids should be kept in 137-237 litres (30-50 gallons) of water. Correctly fed specimens of *Cichlasoma motaguense* and *C. managuense*, for example, would quickly outgrow and pollute smaller water corridors unless massive regular water changes are carried out and excessive filters are brought into play. At 300mm (12in) in length, neither species will achieve a bright, keen-eyed and alert condition if they are confined to a small aquarium.

How can these giants be accommodated? Increasing the width (front to back) measurement of the aquarium is one way of gaining much needed space. This also increases the surface area of the water and thus enhances the rate of gaseous exchange in the aquarium. A superb aquarium would be 240×60×60cm (96×24×24in), but a more realistic system would measure 150×45×38cm (60×18×15in).

For this reason the larger *Cichlasoma dovii* and *Petenia splendida* are much rarer in aquarium circles. Perhaps the attainable size of 500-700mm (20-28in) is a deterrent to all but real enthusiasts and public aquariums. The widespread *Cichla ocellaris* grows to a length of 600-700mm (24-28in) and also requires a great deal of swimming space.

However, bear in mind that the recommended stocking levels cannot be rigid because a great deal depends on the size and strength of the cichlids; even in the perfect aquarium a large sexually active pair of these cichlids can dominate 30-50 per cent of the system!

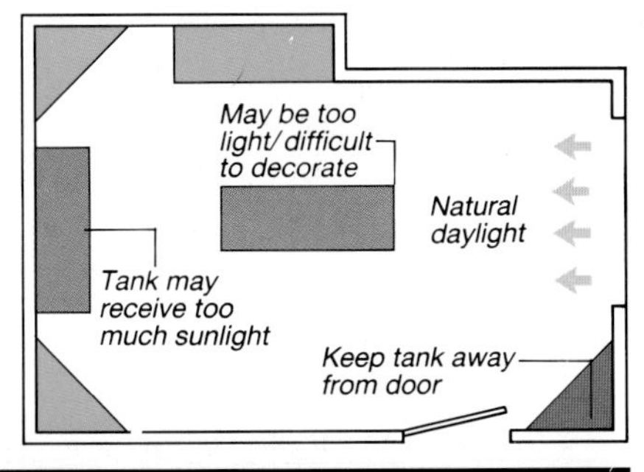

Right: *In this plan, good room positions are shown in blue, reasonable sites are in purple, unsuitable ones in pink.*

Below: *In a Tanganyikan cichlid tank, boulders and rocks create separate territories and provide spawning sites. Plants add interest and variety.*

Coping with large aquaria

It is clear that the size of the system depends on the choice and the number of specimens to be kept. Some aquarists suggest that the problems of aggression can be solved by overstocking the aquarium, in much the same way that African Rift Valley Lake Cichlid territorial problems can be overcome. This route is fraught with danger, however, for infections can spread alarmingly as a direct result of overcrowding and it is very difficult to stabilize water chemistry and quality in these situations.

Some experienced breeders crowd juveniles together in the knowledge that the situation is a temporary one until a breeding pair can be separated out. This strategy can also be successful in large community systems, but it demands a great deal of attention and time-consuming maintenance to the aquarium.

The majority of fishkeepers wish to include other species of fishes in a theme aquarium containing cichlids, such as catfishes, characins and, in some cases, livebearers. The assumption that the larger the system the more fishes can be kept and the easier it is to maintain, applies only if the fishkeeper can afford to incorporate the required filtration systems. It is a common myth in fishkeeping that large aquaria are easier to maintain than small ones; in large volumes of water, pollution just takes a little longer, but when it arrives the headaches are bigger!

It is also sensible to devote serious thought to how such an aquarium is to be supported. Water weighs one kilogram per litre (10lb/Imp. gallon; 8.4lb/US gallon). Allowing for its own

Below left: *Despite being the smallest Central American Cichlid,* Neetroplus nematopus *has the distinction of being one of the most aggressive species.*

Above: Cichlasoma meeki *is an excellent beginner's species to introduce into a smaller or medium-sized Central American system. It is usually widely available.*

weight, a 200-litre (44 Imp. gallon/53 US gallon) aquarium – the smallest recommended quarters for a community of Lake Malawi cichlids – represents just under 205kg (450lbs) of concentrated weight filled with water alone. Few pieces of furniture can support such massive objects. In addition, there is always the small, but real, possibility of furniture being damaged by a leak or by the spilling and splashing of water that is part and parcel of routine tank maintenance. Buying a commercial stand for any aquarium in excess of 80 litres (18 Imp. gallons/21 US gallons) capacity is therefore the most practical solution to this problem. Metal or wood aquarium stands are available in a wide selection of styles, so it is quite easy to find one to complement the decor of any room in the house.

The realities of keeping large cichlids should not and will not detract aquarists from accepting the challenge they present. The brilliance of a male *Cichlasoma* or *Satanoperca* in full display colour and posture is the making of any large aquarium.

Smaller species

Apart from the giants or Guapotes, Central and South American cichlids divide up reasonably well into small- and medium-sized species. The small to medium-sized species can be easily maintained in standard sized aquaria.

The smallest Central American cichlid is *Neetroplus nematopus*, a dwarf species 75-100mm (3-4in) long that is a common, rather secretive inhabitant of the rocky crater lakes of Nicaragua and the Atlantic slope rivers of Costa Rica. Despite its small size, however, it is well able to hold its own among larger cichlids. A dominant male, for example, can destroy the other fishes in a community aquarium and bully cichlids many times its own size. Such aggressive behaviour is a contradiction to the hard and fast rule that directly equates fish size to aquarium size.

Four to six juvenile specimens of *Neetroplus nematopus* could be kept successfully in a suitably aquascaped and filtered aquarium measuring 60×38×30cm (24×15×12in). (Aquarium measurements refer to length, depth and width respectively.) However, it is quite likely that these four or six fishes will eventually finish up as one or, if luck will have it, a spawning pair. In the larger arena of a crater lake or river, beaten off specimens can retreat to safety, but in the confines of an aquarium they have no escape and will almost certainly be killed by the more aggressive fishes. Housing the same population in an aquarium measuring 90×45×30cm (36×18×12in) would be preferable because it would allow fishes lower down in the 'pecking order' greater opportunities to find sanctuary.

The Firemouth Cichlid, *Cichlasoma meeki* and the Tricolor Cichlid, *C. salvini*, could be kept in similarly sized aquariums, although a larger one measuring 120×45×30cm (48×18×12in) could be of long-term benefit. These medium-sized cichlids (about 150mm/6in in length) are widely available from good aquarium dealers. They can be introduced into communities of larger Central American cichlids, providing the aquarium is spacious enough to accommodate the territorial needs of all the fishes.

Remember that breeding cichlids will disrupt any community aquarium. The joy of success in spawning them is often tempered by the damage the pair causes in defending their eggs and fry. Such fierce parental care is the key to survival in their natural habitat and the fishkeeper should only marvel at it and learn to cope with its consequences in the aquarium.

Cichlid fry can be successfully raised in a community aquarium because parents will defend their offspring. The period of parental care varies among individuals and species however, once the fry begin to wander away from the immediate territory, predation by opportunist fishes will occur.

Great Guapotes

Large cichlid species are more often sold as comparative 'fingerlings' or at best juveniles with a great deal of growing to do. The temptation is to add such cichlids into established modest sized community aquarium and worry about their potential growth when that becomes a problem. This 'putting off' the inevitable can result in water quality and fish health problems.

It is reasonable to grow on juveniles of large species within a cichlid community in the knowledge that a larger aquarium

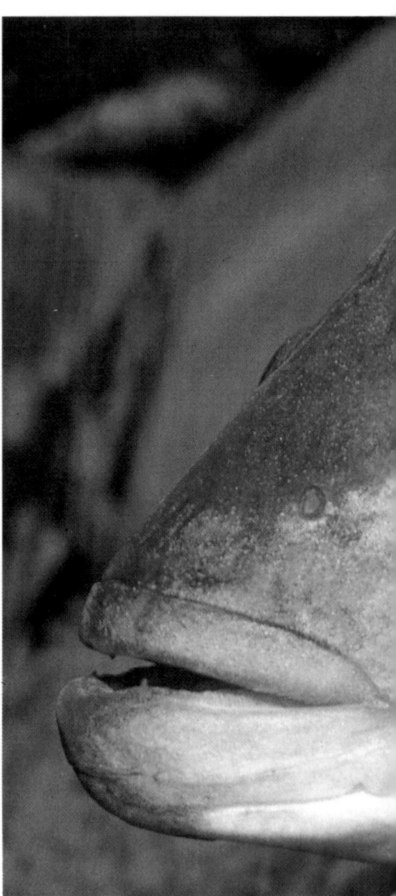

Below: Cichlasoma dovii, *a true giant species that attracts a great deal of interest. Juveniles can be raised easily in small aquarium systems, but adults need very spacious surroundings.*

system is, or will be, established to accommodate their eventual size.

Adding to the established community aquarium

It is never advisable to introduce larger cichlids into small fish communities as those species with large mouths will take every opportunity to attempt to swallow fishes that fit between their jaws. Many experienced aquarists use several 'tricks' to make the introduction successful. They might change the aquascape around or feed tempting foods to distract established fishes when introducing new cichlids. A combination of both ideas will usually be enough to confuse the existing occupants.

If a female cichlid has been spawning in the community aquarium but does not have a mate of the same species it may be courted by other species. This behaviour can happen in captivity but would rarely occur in nature. It can be difficult, or even impossible, to introduce a male of the same species and expect them to be immediately compatible.

There are many stories of newly introduced mates becoming aggressive and bullying the female until it is near to becoming a fatality. It is essential, in an established community or breeding aquarium, that newly introduced cichlids are monitored closely and watched for signs of bullying or territorial behaviour.

Water Requirements and Filtration

Water chemistry

There are two major parameters of water chemistry of importance to aquarists: pH and hardness. The pH of the water is a measure of its relative acidity or alkalinity, or more correctly, the concentration of dissolved hydrogen ions (H^+). The greater the concentration of hydrogen ions, the more acid is the water. The pH scale ranges from 0 (extremely acid) to 14 (extremely alkaline). The pH of pure, distilled water is 7 which is said to be neutral. Because the scale is logarithmic, a decrease of the pH from 7 to a value of 6 is actually a ten-fold increase in the concentration of hydrogen ions. Water pH can be tested using simple colorimetric kits or portable electronic pH meters.

Freshwater fish typically live at a pH level of between 6 and 9, but some South American blackwater species are found in even more acidic water (pH 4-5). It is possible to adjust the pH of water by using chemicals, but such changes should be done very gradually, if at all. Most advanced aquarists avoid changing pH and instead select fishes to suit their own water chemistry.

Water hardness refers to the amount of dissolved salts present in the water. The major salts that affect hardness include calcium, magnesium, carbonate and bicarbonate. Two types of hardness are measured and discussed: general or permanent hardness (GH) which measures calcium/magnesium levels, and carbonate or temporary hardness (KH) which measures carbonate/bicarbonate levels. Carbonate hardness is said to be temporary because it can be boiled off or removed by filtration through peat.

In Europe, water hardness is usually expressed in degrees hardness or dH. One degree of dH is equivalent to 10mg of calcium oxide (CaO) or magnesium oxide (MgO) dissolved in one litre of water. In the USA, the situation is different and hardness is expressed in ppm (parts per million) of total dissolved solids and is equivalent to 1mg of calcium carbonate ($CaCO_3$) dissolved in one litre of water.

Water is classified according to its dH as follows.

dH	$CaCO_3$ ppm	Hardness
0-3	0-50	Soft
3-6	50-100	Moderately soft
6-12	100-200	Slightly hard
12-18	200-300	Moderately hard
18-25	300-450	Hard
>25	>450	Very hard

Water hardness may be measured using commercial chemical 'drop' lather tests, or with an electronic conductivity meter. The conductivity meter measures total dissolved salts and expresses them in microsiemens/cm^3 the more dissolved salts, the better an electrical conductor the water becomes. Although conversion to dH is inaccurate, for aquarium purposes 35 microsiemens is equivalent to 1°dH. Water hardness may be altered to fit the needs of the fish being maintained.

The easiest method to be successful with cichlids is to choose species that are compatible with the mains water supply chemistry. If the tap water is hard and alkaline then most African and Central American cichlids will thrive because the source water is that which suits these groups best. For example, the rivers and crater lakes of Central America and the African Rift are generally hard and alkaline, although some small creeks are acidic and can be more so in the dry season. ('Hard and alkaline' in this case refer to a hardness of 10-30°DH (150-550mg/litre $CaCO_3$) and an average pH value of 8.0.)

If the tap water is naturally soft and acidic (even if the water authorities adjust the water to make it slightly alkaline this can be easily adjusted) then South American cichlids, especially Discus and Dwarf cichlids will thrive. It is more difficult to make hard, alkaline aquarium water into

soft and acidic (reverse osmosis or resin exchanges) than to make soft, acidic water hard and alkaline (simply add bicarbonates and alkaline adjuster or small amounts of crushed coral which can be added into filter medium).

Perhaps more important than water chemistry is water quality – how 'clean' the water is kept. Nitrogenous wastes (ammonia, nitrites, nitrates) which are toxic to fish, build up rapidly in closed aquarium systems. Filtration and water changes provide the solution.

Making Water Hard

As an example, consider a suitably furnished aquarium measuring 90 × 38 × 30cm (36 × 15 × 12in) containing approximately 96 litres (21 gallons) of water with a pH value of 6.9-7.1 and a total hardness of 5-10°dH. Such a system would need about 2.3kg (5lb) of crushed cockleshell added to the substrate to lift the pH value to 7.5-8.1 and approximately 30gm (just over 1oz) of calcium sulphate and magnesium sulphate to increase the hardness to 20-25°dH

Above: *Some coastal species like this Rainbow Eartheater,* Gymnogeophagus rhabdotus *from southern Brazil, appreciate harder, more alkaline water then their Amazonian relatives.*

Below: *A typical cross-section of a creek showing the species community, and distribution. Fishes include cichlids, characoids, rivulins and the predator* Hoplias malabaricus. *After Dr F. Froehlich.*

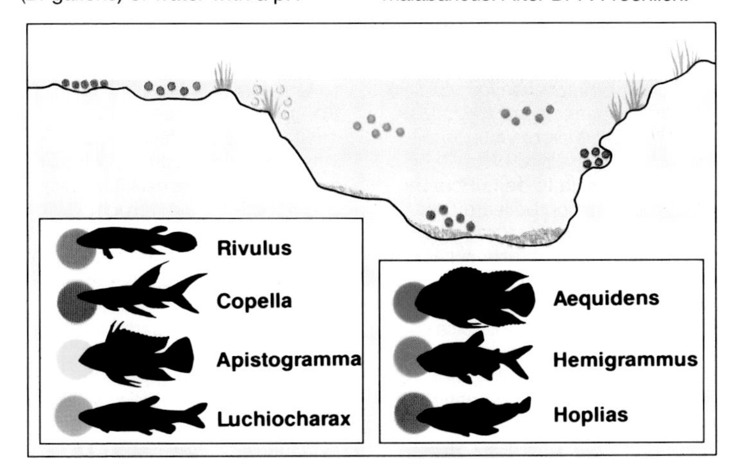

($330-420mg/litre\ CaCO_3$).

The effectiveness of using dolomite chips in the substrate or filter body is continually under review among fishkeepers. Since dolomite – basically calcium and magnesium carbonate – is a hard, fairly insoluble substance, many fishkeepers argue that only dolomite dust can act as a useful pH buffer. Crushed cockleshell or crushed coral may be the better choice for consistent results.

Very soft tap water can be adjusted using commercial salt mixes to create suitable living conditions for Rift Lake cichlids. Used as recommended by the manufacturer, these products will consistently yield satisfactory results. Because they contain a high proportion of carbonates, these mixes take quite some time to dissolve completely – up to three days is not unusual. Take this fact into account, both when setting up an aquarium for the first time and when planning water changes. Remember also, at every water change, to add a quantity of salt mix to the system, equivalent to the amount lost from the tank, in order to maintain pH and hardness at the desired level.

It may be that local tap water has an ideal pH value for keeping certain species of cichlids. In fact, many water authorities treat neutral tap water so that it becomes moderately alkaline, ie with a pH value of 7.5-8.0, to reduce its corrosive effect on water mains. This is an ideal starting point for many African and Central American cichlids.

Whatever type of water is available, it is advisable to keep a constant check on the pH value and hardness of the water and monitor the tap water regularly; both can fluctuate widely.

Making Water Soft

With the exception of species from the coastal rivers, most South American and some African cichlids prefer water whose pH is about neutral or below, and very to moderately soft.

For fishkeepers in hard water areas, there are solutions to make the water soft. One solution is the use of ion exchange resins. These resins (usually two in a mixed bed) exchange sodium (NA^+) for calcium (CA^{++}) and magnesium (Mg^{++}) ions (cation exchanger), and chloride (Cl^-) for carbonate (CO_3^-) and sulphate (SO_4^-) ions (anion exchanger). While they soften the water it is nevertheless 'salty' with Na^+ and CL^-. Moreover, the resins must be regenerated at regular intervals, using hydrochloric acid and sodium hydroxide, corrosive and potentially dangerous chemicals.

A better solution to creating soft water is the use of a reverse osmosis (RO) system. In reverse osmosis processing, tap water is passed under pressure through a membrane, whose pore structure is small enough to allow water, but not dissolved minerals, to pass through. The 'strained' water is

collected into a sump (typically a plastic litter bin or vat). It is essentially distilled and may actually require 'hardening' by the addition of salts. RO processed water allows complete control over the composition of aquarium water and is particularly useful for delicate blackwater species. Membranes must be back-flushed regularly and replaced semi-annually as needed. Although initially expensive, RO units are well worth the investment if you are serious about keeping rare South American cichlids.

Demineralizing water may also be obtained commercially in small volumes as bottled distilled water.

Left: *A chalk barrier in Mexico, home water to several species, including* Cichlasoma intermedium, *a relative newcomer to aquarium circles. Good water flow can be essential to simulate habitat.*

Below: *In keeping with all Central American riverine cichlids,* Cichlasoma intermedium *will thrive in bright water of pH 7.5-8.1.*

When keeping a small number of dwarf cichlids in small volume tanks, this may be a workable solution. However, remember to add some aquarium salt (non-iodized NaCl) at a rate of 1g per 45 litres (1 level teaspoon to 10 gallons). Rainwater is not recommended because of the industrial pollutants and impurities which inevitably occur in it. If none of the above suggestions are possible, then simply select fish that are compatible with local water.

Often the water must be acidified. This is best done naturally using either peat or peat extract. Peat leaches natural humic and tannic acids which acidify water filtered through it. Moreover, it will also chelate toxic heavy metals. Peat can be obtained in bales or in pellets. Make sure it does not contain fungicide or fertilizer. Commercial peat extracts are available in the trade.

Keeping Good Water Quality

Central American and African cichlids appreciate partial water changes of 30-50 per cent every 7-10 days. In a heavily stocked aquarium increase the frequency of the water changes and the amount of water replaced. Malawian cichlids, in particular, seem to relish a regular replacement of up to 85 per cent of their tank's volume. However, Tanganyikan cichlids respond less favourably to large-scale water changes. Representatives of the genus *Tropheus* and its allies appear the most tolerant of such a programme, whereas *Lamprologus* and representatives of allied genera, seem more sensitive to such practices and usually refuse to breed when managed in this way. Fortunately, most of these Tanganyikan cichlids will thrive as single pairs and the waste load produced in such a lightly stocked tank is unlikely to strain the capacity of a well-established biological filter.

In nature, South American cichlids inhabit light brown to dark brown waters with a pH range 7.2-6.0 with a general hardness range between 12 and zero degrees. Aquarists living in an area where the water supplies are naturally soft and acidic will be successful with most South American cichlids. Some species from coastal lagoons and creeks may live in water above these ranges but carbonate and alkaline adjusters will quickly make the source water acceptable.

Below: *This graph shows the chemical changes that occur in the early life of a new aquarium. Within two or three weeks of fishes being introduced, the level of ammonia (A) in the water peaks. As this falls, the nitrite level (B) reaches a maximum within three to five weeks. The level of nitrate (C) builds up over several months and is reduced by regular partial water changes. Seeding new filters with mature media reduces these chemical fluctuations.*

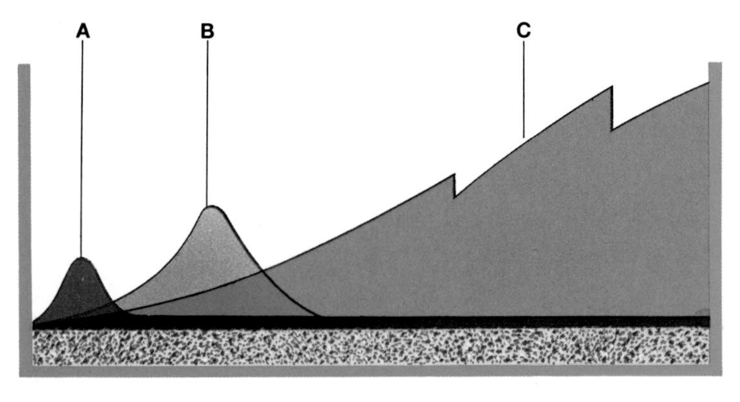

Experience shows that simple partial water changes often have no lasting effect on raising the pH value of tank water that has become undesirably low. For example, in an established tank with a pH value of 6.8, performing a 30 per cent water change using a domestic water supply of pH 7.2 will still leave the tank pH value at around 6.8 or 6.9. Adding a pH corrector, such as sodium carbonate, will temporarily lift the pH value, but this effect will not last. However, extracting silt from the gravel and making a 30 per cent water change at the same time, plus the addition of a pH corrector, will have the positive effect of lifting the pH value to the ideal region of 7.5-8.1.

The pH value of water in the enclosed system of a tropical aquarium is influenced by many factors, including the substrate, the amount of organic debris and the number of fish the tank contains. More fishes equals more waste and excess protein in the system, which leads to decreasing oxygen levels, a rise in carbon dioxide and a consequent fall in pH value (ie the water tends to become more acidic). The fish show distress by increasing gill rates as they attempt to extract decreasing oxygen from the water in the aquarium.

Monitoring the pH value of the water regularly will reveal such trends, but more direct visual clues will help to judge the water quality. Look for bubbles clustering on the water surface as a telltale sign of excess protein, which increases surface tension. Draw off a glass of tank water and stand it in front of a white card; if it is clearly yellow then consider making a partial water change and cleaning the substrate of organic debris. If high nitrates are recorded, increase the percentage and frequency of partial fresh water changes.

Achieving a stable pH is the long-term target. Aim for it by monitoring the water regularly, by avoiding overstocking and by being generally aware of the system in your care.

Water Movement

With the exception of specialized species that live among the rapids, such as *Steatocranus casuarius, Teleogramma brichardi* and *Haplochromis bakongo*, African cichlids do not relish strong water movement. Most riverine species seek out areas of moderate water flow and, apart from species that inhabit the surge zone, lake cichlids are also found in areas of relative calm. In captivity, African cichlids find the degree of water movement provided by a working filter satisfactory.

South American cichlids vary in their water movement needs

Below: *This combination of undergravel and power filtration is ideal for Central American cichlids. Each system works independently. A venturi adds beneficial air to the water outlet from the power filter.*

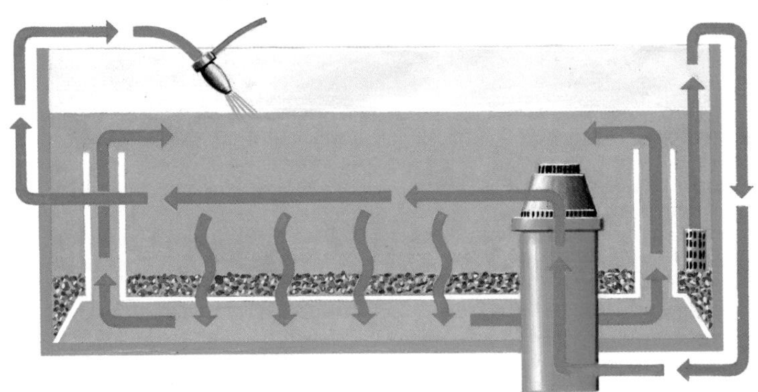

although aeration is usually essential. Discus and smaller cichlids prefer very little aquarium water movement and will avoid strong filter or pump flows except to seek out freshly introduced food.

Other than in very densely populated aquariums, the return flow of an outside power or canister filter creates sufficient surface turbulence to facilitate the free exchange of gases from solution. However, filters have been known to suffer temporary failure if they become blocked, or if the siphon action is interrupted. The respiratory needs of African cichlids are proportional to their size, and cichlid communities are often densely populated. The consequences of filter failure can be swift and irreversible, so it is worthwhile including supplementary aeration in a tank. A small diaphragm pump and an airstone will suffice to ensure against possible disaster.

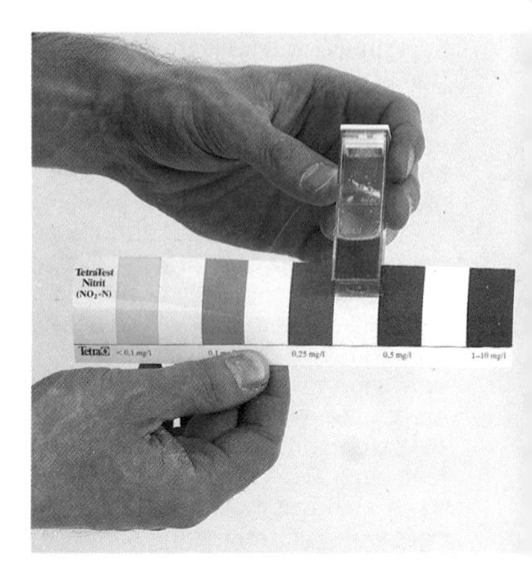

Above: *Monitoring nitrite levels in aquariums, especially new ones is vital. Here, a sample or tank water treated with a reagent is compared to a nitrite level colour chart. Such test kits are available for checking pH value and other parameters.*

Ammonia, nitrite and nitrate

The major consideration for newly established African and Central American cichlid aquaria is that ammonia is a greater poison in alkaline water – ie within the range of pH values that are ideal for these fish – than it is in water of a lower pH value. This factor alone has caused more fish deaths than any published statistics will ever show. (Acidic water contains a higher concentration of hydrogen ions (H+) than alkaline water and these combine with the toxic ammonia gas (NH_3) to form less harmful ammonium ions (NH_4^+). In water with a pH value of 8.0, however, five per cent of the ammonia/ammonium content is present in the toxic ammonia form; at pH9 the level of free ammonia increases five fold.)

In the normal course of events, the first few weeks of a new aquarium's life are critical. If the filtration system has not been seeded with gravel or filter medium taken from a mature system or a commercial bacteria and enzyme kick start, the prospect of ammonia and then nitrite poisoning is an ever-present danger. Ammonia, a direct result of the fish's waste products, peaks first. As the activity of nitrifying bacteria increases, much of the ammonia is oxidized to nitrite (NO_2^-), which produces a second peak. The natural process of nitrification – a process encouraged in filtration systems, converts much of the nitrite to the less harmful nitrate (NO_3^-).

Once a system has been established for several months, ammonia and nitrite levels should no longer pose a threat. Nevertheless, it is wise to make water tests on a regular basis to ensure nitrification remains stable. Cleaning the filtration media and making regular partial water changes will normally keep the nitrate levels within acceptable limits.

The presence of high nitrates, often a direct result of increased stocking levels or size growth of the cichlid community, requires an immediate increase in the percentage and frequency of partial fresh water changes. It would also be relevant to review stocking levels and to investigate if a noticeable level of nitrates are also in the tap water supply, as these factors can also be dealt with by a reduction of fish stocks and denitrate filters, or the use of a specific nitrate removing resin in the filter.

Large partial water changes are not only beneficial to fish health but can often trigger cichlids into spawning behaviour. Cichlid breeders will use this simple stimulus on almost all the range of species available including African lake Mbuna and Discus.

Filtration

Large healthy fish fed correctly produce plenty of waste. In the wild, waste products are diluted immediately in relatively large (and usually moving) volumes of water, and the natural process of nitrification (ie the conversion of toxic ammonium compounds to less harmful nitrates by bacterial action) continually purifies the water. In the closed environment of an aquarium, however, filtration and aeration are necessary parts of the basic life support system for the fish. To spend a higher percentage of the total budget on aquarium decoration than on filtration would be a serious error. Sponge filters, either air driven, or motorized, are an excellent choice for any situation where a minimal waste load is to be treated. The extreme porosity of the sponge material allows it to support a bacterial community comparable to that of an extensive gravel bed. As long as there is no mechanical blockage of the sponge's pores to obstruct the free flow of oxygen-laden water through the sponge medium, such a filter will break down metabolic wastes efficiently.

Such filters are particularly well suited to aquaria for the smaller substrate-spawning species from Lake Tanganyika, as well as for many dwarf and middle-sized

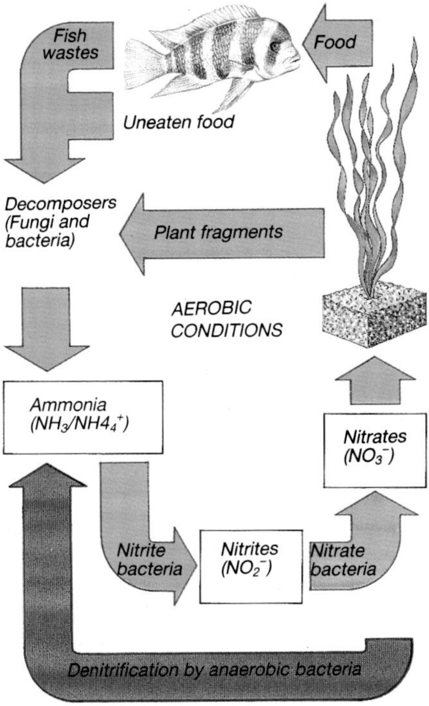

The nitrogen cycle

The natural processes of the nitrogen cycle convert the dangerous ammonia formed by the decomposition of fish wastes and uneaten food into less toxic products, such as nitrates, which are used as a plant food. Nitrifying bacteria (such as Nitrosomonas *sp.) combine the ammonia with oxygen in the water to form slightly less toxic nitrites. Further bacteria (such as* Nitrobacter *sp.) continue the cycle; with the addition of further oxygen, nitrites are converted into even less harmful nitrates. A plentiful supply of oxygen in the water is essential if the helpful bacteria are to survive and multiply. Provide adequate aeration and never turn off filters for long periods.*

riverine cichlids. Air-driven sponge filters are also suitable for nursery tanks for female mouthbrooders, and in fry-rearing tanks. Air-driven units combine efficient operation with complete mobility and are available in a full range of sizes. The smaller units have the further advantage of being relatively unobtrusive and easily camouflaged with rocks or plants.

Internal power filters, incorporating a bioactive sponge medium, are also readily available. These units are particularly useful in larger tanks, where strong water circulation is required. However, their motors generate waste heat and rely on the tank's water for cooling. In this case, the filter capacity must match the tank volume, to prevent overheating, particularly during the summer. Fortunately, internal power filters are available in a number of different sizes, so it is a simple matter to match the size of the unit to the volume of the tank.

Smaller internal filters do not perform satisfactorily in an environment where the waste load is so heavy that it actually clogs the pores and obstructs water movement. Such a situation arises in tanks housing very large cichlids, such as the various tilapias, or in those that are very heavily stocked, such as Malawian cichlid communities. In these cases, the most efficient filtration strategy is to combine conventional undergravel and power filtration in parallel one system acting as an independent back-up to the other.

It is vital to use at least 7.5cm (3in) of rounded gravel over the filter plates. A particle size of 5mm (¼in) is ideal. Place rocks around the base of the uplifts to prevent them from being displaced by digging or fighting cichlids. Also, place a gravel tidy (simply a plastic mesh) in the substrate to prevent digging cichlids from uncovering the filter plate.

Power heads (electrically driven water pumps) can be fixed to the top of the uplift tubes to speed the

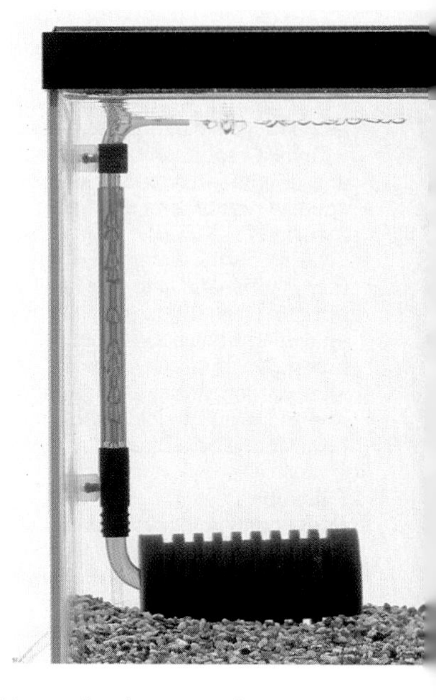

Above: *An air-operated sponge filter is ideal for the cichlid tank with a minimal waste load, and in nursery and fry-rearing tanks.*

flow of water through the undergravel filter. Some power heads aerate the water only if they are positioned close to the water surface, so adjust the uplift tubes if necessary to produce this beneficial aeration. Most power filters have venturi adaptors to allow air to bleed into the water flow and this also enhances the level of oxygen in the water.

In order to provide additional and reliable nitrification over a long period, load external power filters with long-term filter media. The alternatives of filter carbon and sponge medium will both support nitrifying bacteria but need more frequent changing. Filter carbon has been advocated for many years but its main disadvantage is that its lifespan cannot be predicted. It has an incredible surface area for absorbing waste materials, but once this is saturated, organic waste can leak back into the system. The effectiveness of most remedies is

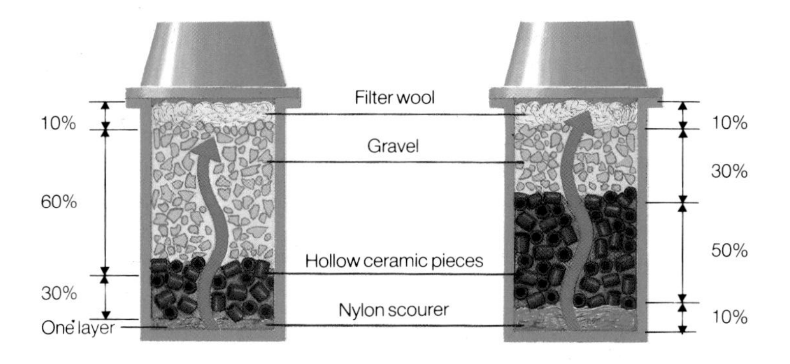

Above: *Recommended media for power filter canisters to ensure good water flow and efficient filtration.*

reduced by active carbon and it should be removed before treatments are introduced into the aquarium. Sponge medium is more or less useless because once the surface area becomes greased with organic slime, water tends to flow over it rather than through it.

Although undergravel and power filtration make an ideal combination, it is possible to use either in isolation. In this case, be even more rigorous in maintenance routine. For aquaria operating purely on undergravel filtration, be sure to remove sediment from the gravel during weekly or fortnightly partial water changes. To operate an aquarium solely on power filtration, ensure that extremely powerful canister filters are utilized and that the substrate in the aquarium is shallow. A light scattering of river sand or a shallow pebble layer would be ideal, but this will need to be raked through at least once a week.

Ideally, an aquarium housing cichlids should provide both still and moving water, allowing the occupants an element of choice. In the wild, a single widespread species may well be recorded in both types of habitat. For the fishkeeper, it is easier to establish such distinct regions in a large aquarium rather than in a small one. A possible strategy is to install a power filter or power head at one end of the aquarium and a gentle aerated uplift at the other. A flow would be generated around the aquarium but a pocket of relatively 'quiet' water would also develop.

Perfect though this sounds, it could lead to unexpected 'social' problems within the aquarium. Strong fish or those at the top of the pecking order may bully those fish that prefer to remain in the slow water areas. Breeding pairs, for example, would invariably choose the shelter of such slow water in which to spawn. Thus, in defending such ideal territory, unfortunate low pecking order fish would be an easy target for excessive violence.

Ideally, layer the inlet/base area of the power filter with a nylon scourer and some ceramic hollow pots. This will allow water to pass through virtually unimpeded but will trap larger particles and organic debris. Fill the upper part of the filter canister with porous gravel as used in the aquarium. This will act both as a physical and as a biological filter once bacteria become established on the surface of the gravel particles. Finally, fill the upper part of the filter body with a 5-7.5cm (2-3in) layer of filter wool to extract fine particles of sediment from the water as it leaves the filter.

If the aquarium contains large cichlids, reduce the amount of gravel substrate and increase the proportion of ceramic pots in the filter body. If necessary, increase the water flow by using a coarser gravel.

Two high-tech solutions have revolutionized fishkeeping: the canister filter and the wet/dry trickle filter.

The canister filter is an external power filter whose filtration media are contained in a cylindrical sealed canister. The media include floss (mechanical) ceramic rings or foam block. A pump pulls water from the tank through the canister and returns it via a spray bar.

Wet/dry trickle filters, originally designed for marine tanks, have proven effective in freshwater tanks. These are also external, power-driven filters. The filter consists of a large acrylic box into which are stacked porous plastic balls, cubes, etc. which provide the colonization bed for the nitrifying bacteria. These plastic media have a tremendous surface area and lots of hollow spaces that enable water to 'trickle' through them. The water from the tank, mechanically pre-filtered, is delivered via a rotating spray bar or a perforated drip plate to the stacked media and allowed to trickle through it (the 'dry' phase). In this aerobic environment, ammonia is detoxified and the water oxygenated. The trickled water collects in a second compartment, often passing through a foam filter, and is pumped back to the tank via a submersible pump.

When installing a new filter, always 'seed' the nitrification process by adding some mature gravel or filter medium (ie used in an aquarium over 8-10 weeks old) to the new medium. A good aquarium dealer will provide suitable material from an existing aquarium.

Selecting the correct filter size

It is impossible to over-filter aquarium water; clean water is to fish as clean air is to humans.

For any given aquarium capacity a power filter should be capable of turning over that volume two to three times per hour *under pressure*. Manufacturers tend to quote turnover figures for their filters based on an unloaded canister under no pressure. Once sediment is drawn into the filter medium, the flow rate can fall drastically.

If your budget cannot stretch to an external power filter when you set up the aquarium, then all is not lost. An undergravel filter (once it has established the bacterial colony) can cope with a new aquarium system for 8-12 weeks. To spread the costs, add a power filter to the system at this stage.

Remember, there is no one filtration approach that is perfect. In fact, most successful aquarists use several filter types on each

A thicket of plants provides hiding places favoured by these shy fish. Java ferns and floating plants are excellent choices.

Above: *A typical set-up for Dwarf Cichlids or medium sized acaras includes both live plants, spawning caves and bogwood shelter.*

large tank. For instance, simple box filters providing mechanical filtration and aeration coupled with external power, or canister filters providing chemical/biological filtration.

Whatever the system employed it is essential to back it up with regular water changes. It is possible to start a cichlid community with juvenile or small fishes and upgrade or add filters as the fishes grow. Internal and external filters are easy to upgrade as funds allow and the use of existing filters can be continued until new filters have become organically mature. Even the best filters, whatever the type, will not cope with overstocking and poor water quality resulting from inadequate partial water changes. Each community system has its own filter requirements, dependent on the number of fishes and the size of the aquarium. It would be very difficult to over filter a system and therefore it is ádvisable to purchase equipment that has a rating for a larger system than the one for which it is intended. This advice comes in handy when fish health problems occur in the aquarium as a higher standard of water quality can be guaranteed. Often one type of filter can be used to back up another and means that one filter can be cleaned whilst the other remains stable. This is especially true for internal and external power filters are used in tandem but also applies to air or power generated plates and internal or external filter combinations.

Pieces of cured bogwood provide shelter for these shy fish as well as a visual counterpiece for the aquarium tank.

The use of undergravel filtration should be limited to small cichlids like apistos, which produce few wastes and do not dig.

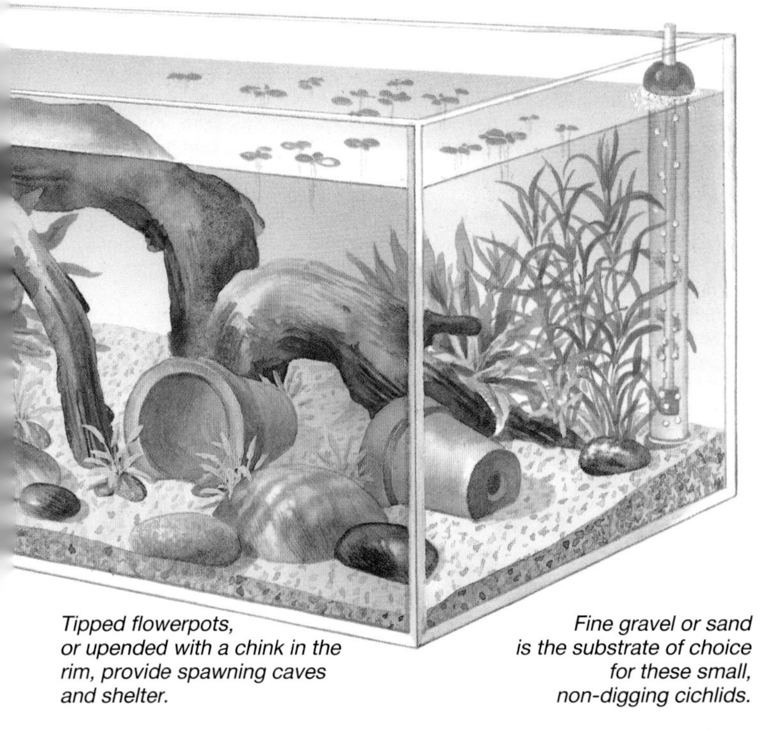

Tipped flowerpots, or upended with a chink in the rim, provide spawning caves and shelter.

Fine gravel or sand is the substrate of choice for these small, non-digging cichlids.

Right: *An external power filter. Water drawn from just above the gravel bed is cleaned and returned to the aquarium through a horizontal spray bar that runs the length of the tank. Electrically operated units provide an increased water flow compared to air-driven filters. Add an appropriate filter medium (e.g. a sponge block or net bag of ceramic rings). Always keep filters clean. In large tanks, combine a power filter with a motorized internal filter for best effect.*

Stagnating substrate

If internal or external power filters are in place and undergravel filters are not employed it is essential that the substrate level is established no deeper than a few centimetres. This is best achieved with a shallow scattering of river sand. A deep substrate can quickly stagnate and will rob much needed oxygen levels from the water. The bacteria forming during stagnation can infect substrate haunting species, especially those that enjoy sifting through the gravel or sand. The most visible signs indicating pollution and stagnation in the substrate is fishes with a swollen abdomen, heavy respiration and unusual upper water swimming. Many unexplained cichlid deaths, especially to rams and other dwarf cichlids, are undoubtedly linked to substrate stagnation. Few, if any, commercial books on keeping cichlids have over pointed out the problems resulting from unfiltered substrate.

It is wise to consider which species are to be kept before the filter type, aquarium size and the aquascape are decided upon. Small, non-substrate sifting, cichlids can easily be maintained in an aquarium filtered via an undergravel plate. The uplifts, operated by air pump and power head, displace water which is then replaced through the filter plate. This system does not permit sand to be used as this fine substrate, when used at the necessary depth of 10cm, quickly packs when introduced on to an undergravel filter plate. The alternative, when a sand substrate is preferred, is to use one of the many types of internal or external power filters that are available on the market today.

Medium to large cichlids are well known as messy dry food feeders. They often take in pellets, flaked food or food sticks and expel clouds of crushed-up food from their gills. In communities of mixed fish sizes some of this waste food is welcomed by smaller species. If there are no 'sweepers up' of food in the cichlid aquarium the food waste is quickly drawn into adequate filters and can lead to a reduction in flow rates and lowering of water quality. Larger power filters, even though powerful and slower to lose flow rates, should still be cleaned on a regular monthly basis to clean away excess food.

Water changing

Remember to disconnect the tank's heater when making a water change. If the water level drops too far, the heater tube will seriously overheat and almost certainly crack when it comes into sudden contact with cooler replacement water. Many experienced aquarists also believe that it is not advisable to combine large-scale water changes with the replacement or cleaning of the tank's filter medium. Insofar as the presence of organic matter tends to neutralize free chlorine, there is some truth in this as far as a mechanical unit is concerned. However, the function of a biological filter will not be affected if the medium is rinsed in warm, rather than hot tapwater.

Most of the mess and bother associated with replacing water are eliminated if one uses an automatic water changer. This apparatus – part siphon-powered gravel cleaner, part water bed pump – attaches directly to the tap and, with a flip of a control knob, changes from a powered siphon to a hose. Thus you can combine a water change with the removal of waste from the substrate, another important aspect of nitrogen cycle management.

Before using raw water for water changes, it is a good idea to check its chemistry. In these days of acid rain, metropolitan water sources frequently 'lime' the water near the pumping source making it more alkaline so delivery pipes will not be harmed by its natural acidity. What comes from your tap may measure pH 7.9, but if it is very soft, it will have no buffering capacity and the pH will drop, harming your fish. You may have to draw water first into a reservoir and allow it to stand, or filter it with a biological (eg sponge) filter before use.

Breeders may find the action of an automatic water changer a bit too powerful in a tank containing newly mobile fry. It is a good idea to collect the waste water from a fry tank in a bucket before pouring it away, thus allowing any wayward fishes to be rescued from the waste water and returned to safety of their tank.

Below: *Many of the larger cichlids like this* Aequidens tetramerus *are heavy and messy eaters and require the power and efficiency of mechanical filtration to remove the suspended particles of uneaten food and waste they produce. Make sure to clean the filters on a regular basis to ensure that particles of waste are properly removed from the system.*

Heating and Lighting

The great aspect of cichlids is that they are mainly tough and adaptable. Few hard and fast rules can be applied to them although it can be helpful to examine the natural habitat of certain cichlid groups to understand about the type of environment they have adapted within.

The majority of African cichlids are fish of the lowland tropics. While many riverine species regularly encounter temperatures as high as 35°C (95°F) in nature, no cichlid found in such habitats ever experiences temperatures lower than 18°C (64°F). Rivers flowing through open savanna may experience daily temperature fluctuations of up to 7°C (13°F) during the dry season, but the water temperature of streams flowing under forest cover may vary by no more than 1-2°C (2-5°F) over a 24-hour period. Inshore habitats in the Great Lakes are virtually devoid of significant diurnal temperature variation and vary no more than 1-2°C (2-5°F) on a seasonal basis.

Likewise, in the wild, Central American cichlids seem able to thrive in a wide range of temperatures. Hans Mayland, a West German aquarist, has collected cichlids in various countries. In the Panamanian rivers of Capira, Mendoza, Chagres and Tupisa he found cichlids flourishing in recorded water temperatures of 25.6-31.2°C (78.1-88.2°F). (Air temperatures were in the range 22-23°C/72-91°F.)

In captivity, most cichlids prosper over a temperature range of 21-29°C (70-85°F). Aquarists prefer to maintain them at the upper end of this range – in fact Central American cichlids prefer a temperature range of 25-29°C (78-85°F) – on the grounds that spawning occurs more readily. In the case of mouthbrooding species, an added benefit is that the incubation period is shorter at higher temperatures. On the other hand, cichlids are significantly more aggressive at the upper end of their preferred temperature

range. Their metabolism is also accelerated, leading to heftier appetites and an increased waste load for the aquarist to cope with. There is also reason to believe that the lifespan of dwarf forest-dwelling riverine cichlids and of the popular mbuna of Lake Malawi are shortened by continual exposure to higher temperatures. The advantages of maintaining African cichlids between 21-23°C (70-75°F) outweigh any disadvantages. To trigger spawning, increase the temperature by a few degrees.

Most Neotropical cichlids, on the other hand, should be kept in the range of 25-29°C (78-85°F). Discus, Angel fishes and Dwarf cichlids thrive in the upper part of this range, which means that the oxygen content of the water will be relatively low (there is more dissolved oxygen in cold water than in warm).

Cichlids are able to tolerate changes of 1-2°C (2-5°F) in either direction over a period of 45 minutes to an hour. To some extent this simplifies making water changes, but more extreme changes may provoke an outbreak of 'ich'. Tanganyikan cichlids, in particular, are extremely sensitive to abrupt temperature fluctuations, particularly downwards. A reliable thermometer is an essential piece of equipment but unfortunately, precision is not a feature of the thermometers manufactured for aquarium use. To overcome this problem, invest in an accurate darkroom thermometer, and use it to calibrate the tank thermometer.

A reliable, thermostatically controlled heater is vital. Units with submersible heater elements are particularly efficient in large tanks, where they can be sited to take advantage of warm water's tendency to rise. This greatly simplifies the task of ensuring that the aquarium is uniformly heated. However, any well designed and constructed unit can be successfully incorporated into an African cichlid tank.

Heaters that use solid-state thermostats although more expensive, are far less vulnerable to failure than are units that rely on old-fashioned bimetallic strip thermostats. Always match the heater output carefully to the actual tank volume. Calculate the volume of water in the tank by subtracting from the rated tank volume the space taken up by the gravel and furnishings. Alternatively, keep a note of the quantity of water used to fill the tank for the first time, after setting it up. A heater's output is measured in watts. The higher the rated wattage of a unit, the greater its output of heat. Allow 10 watts/4 litres (approximately a gallon) of actual tank volume. Limiting the heater's wattage to a sensible maximum guarantees that the water in the tank will never be heated beyond a temperature that may prove lethal to the stock.

The table shows the recommended number and power ratings for various aquarium sizes. If in doubt, ask your local dealer for advice.

Above: *African cichlids prosper in bright or dim lighting conditions. For plant growth, combine warm white and colour enhancing tubes.*

To boost the oxygen levels for cichlids in warmer waters, employ heavy aeration with multiple air stones or internal air-driven filters. This will aid in mixing the oxygenated surface water, thus increasing the general oxygen content of the water. A wet-dry trickle filter can also be used in combination with aeration to boost oxygen levels. This is particularly helpful when keeping rheophilic (rapids-loving) species which depend on higher dissolved oxygen contents.

Lighting

Lighting an aquarium successfully can create a marvellous picture. On the other hand, a badly lit aquarium appears dull and unimaginative. Choosing the appropriate method of lighting is thus an important decision for both the fish and onlookers alike.

Fluorescent tubes are the norm for aquarium lighting. They are cheap to run and cool in use, and usually serve their purpose quite satisfactorily. However, they direct an even strip of light into the aquarium that produces a predictable effect. The alternative is to use tungsten spotlighting. This can create truly dramatic effects but for the difficulty of accommodating the lamp fittings into a narrow aquarium canopy.

Nor is it possible to use them in

conjunction with glass covers. Even placed well above the water surface, they generate a fair amount of heat which can complicate the task of maintaining a constant tank temperature.

Using both types in combination – an option favoured in West Germany and Holland – is an excellent choice, especially when such combinations include fluorescent tubes that produce a balanced light output for encouraging plants to grow in aquarium conditions.

If vigorous plant growth is required in the African or South American cichlid tank, illuminate it at the rate of 1 watt per 4 litres (approximately a gallon) of water, using either a combination of warm white and colour enhancing bulbs or full spectrum fluorescent tubes, for 14 hours per day. This formula assumes a tank depth of 30cm (12in). For every additional 8cm (3¼in) of tank depth, double the wattage value in the above formula. As rated wattages increase in 5- and 10-watt increments, this is only a guide and it is not always possible to install a lighting system that corresponds exactly to the value of the calculations. In such instances, always round the actual wattage of the tubes upwards to the closest possible value. If excessive algal growth should occur, gradually cut back the number of hours that the tank is lit each day. Often the only way to bring an algal bloom under control is to reduce the number of fish in the tank.

Since plant growth represents a very low priority in Central American cichlid systems, experiment instead with spotlamps to achieve dramatic lighting effects.

Most African cichlids are indifferent to light intensity and can be expected to do well in brightly lit tanks. The smaller forest-dwelling riverine species and a number of cichlids found at depths in excess of 3m (10ft) in the Great Lakes prefer more diffuse illumination in captivity. A screen of floating plants on the surface of their tanks is enough to satisfy them.

Lighting effects in the aquarium are for your enjoyment as much as for the fish's benefit. For best affect time them to coincide with your fish-watching activities.

Below: *This 160×60×75cm (63×24×30in) aquarium is lit by two 65 watt Grolux fluorescents at the front and two 100 watt spotlights.*

Using Spotlights

In deep-water aquarium systems, where spotlighting can be much more effective than fluorescent tube lights, the standard canopy can be dispensed with altogether. The lights can be hung over the open top aquarium and controlled, for height and photoperiods, separately.

Suspend the lights in attractive fittings or conceal them behind a facade. Once installed, powerful spotlights directed at a rippling water surface will produce a striking display of angled shafts of light in the aquarium. The cichlids will quite happily swim in and out of these lighted areas.

Whatever type of lighting is used be sure to keep the aquarium regularly maintained so that the lighting is used to its maximum effect. An excess of tannic acid in the water – leaching from unsoaked bogwood, for example – will cause a yellow-brown cast that severely reduces the penetration of light. Efficient filtration and regular partial water changes should prevent any major discoloration of the water. It is also vital to keep the cover glass clean to allow as much light as possible to reach the aquarium.

Recommended lighting systems for plant and algal growth

Aquarium size (L × D × W)

60 × 38 × 30cm (24 × 15 × 12in)
1 × 45cm (18in) specialist aquarium fluorescent tube (15 watt) 12 hours per day
or
1 × 60 watt spotlight 6 hours per day

---

90 × 45 × 30cm (36 × 18 × 12in)
1 × 75cm (30in) specialist aquarium fluorescent tube (25 watt) 12 hours per day
or
1 × 75 watt spotlight 6 hours per day

---

120 × 45 × 38cm (48 × 18 × 15in)
2 × 90cm (36in) specialist aquarium fluorescent tube (30 watt) 12 hours per day
or
2 × 75 watt spotlights 6 hours per day
or
1 × 90cm (36in) specialist aquarium fluorescent tube (30 watt) 12 hours per day combined with
1 × 75 watt spotlight 6 hours per day

---

150 × 45 × 38cm (60 × 18 × 15in)
2 × 120cm (48in) specialist aquarium fluorescent tube (40 watt) 12 hours per day
or
3 × 75 watt spotlights 6 hours per day
or
1 × 120cm (48in) specialist aquarium fluorescent tube (40 watt) 12 hours per day combined with
2 × 75 watt spotlights 6 hours per day

If plant and algal growth is not required, lighting times can be reduced by 50%.

Recommended heating systems

Solid state Electronic heaters.

Aquarium size (L × D × W)

60 × 38 × 30cm (24 ×15 × 12in)
1 × 100 watt heater-thermostat

---

90 × 45 × 30cm (36 × 18 × 12in)
2 × 200 watt heater-thermostats

---

120 × 45 × 38cm (48 × 18 × 12in)
2 × 200 watt heater-thermostats

---

150 × 45 × 38cm (60 × 18 × 15in)
3 × 200 watt heater-thermostats

(The recommended heater ratings are slightly higher than are normally suggested, but experience shows that this provides a vital safety margin.)

Aquascaping

Aquascaping can provide a stable and easily maintained environment and an aesthetically pleasing 'picture' for onlookers to enjoy.

Recreating the fish's natural habitat is one of the greatest challenges that face fishkeepers and is often the priority for all animal keepers. Working within reasonable parameters, it is possible to provide a cichlid community with an environment which, although not quite in the natural dimension, parallels the real situation sufficiently to sustain the aquarium occupants. The beauty of an aquascaped aquarium is very much in the eye of the beholder, but most people – even non-fishkeepers – would agree that a successful result can enhance a living room and provide an endless source of conversation.

Of course, behind the 'scenes' of any aquarium layout lie the essential service of filtration, aeration, lighting and heating.

Fish in general, and cichlids in particular, will live and breed happily in aquarium surroundings that bear very little resemblance to their native habitat. However, to the observer, ornamental fish unquestionably make a more satisfactory showing in a reasonably naturalistic setting, with the proviso that the biotopes for example of most African cichlids cannot be duplicated exactly in captivity. Nor, in the majority of instances, would aquarists find the results of such an exercise to their liking if they could. Aquascaping a cichlid tank is like gardening; its object is to create a setting that shows the fish to best advantage, while satisfying their need for shelter and breeding sites.

Selecting the correct substrate
The substrate dictates the overall atmosphere of any aquarium. In the wild, Central and South American cichlids swim above pebbles, leaf-littered mud or sand. Large pebble gravel can look fantastic in an aquarium but allows uneaten food to become trapped and to pollute the water. River sand, superb when rippled by a water flow, can become packed and starved of oxygen. This can lead to adverse bacterial activity and serious infections spreading among the fish in the aquarium.

External power filtration can maintain the water quality in aquariums using either of these substrates. If undergravel and power filters are operated in conjunction, both correctly established and maintained, then pebble gravel or river sand are ideal and safe.

The boundaries of a fish's territory may be closely linked to the pattern of the substrate. A cluster of four or five small boulders, for example, may form the dividing line between one territory and the next. In a crater lake, a piece of driftwood may act as a territorial 'marker'; in a river it could well be a cobbled area or a fallen tree. Territorial boundaries also depend on the size of the cichlids in question and the number of fish in a population. For dwarf varieties which require planted tanks and do not dig, fine gravel or even sand is appropriate: a substrate that will anchor rooted plants but will not pack unduly. This same substrate is also appropriate for the sand-sifting Eartheaters (*Geophagus* species). For large, gravel-excavating species, bare tanks may provide the best, most hygienic, though less aesthetic, solution.

Take into account the chemical composition of the aquarium substrate, since this will in turn affect the water chemistry in the tank. Any material intended for use in a tank housing soft, acid-water species must either be chemically inert or have an acidic reaction, lest it raise pH and hardness values to unacceptable levels. Pure silica substrata, such as cage bird grit or fine silica sand, are chemically inert. So are quartz and flint gravels and epoxy coated materials. Laterite and many basalt-derived gravels, on the other hand, will slightly acidify aquarium water.

To test for soluble minerals capable of raising pH and hardness, place a small sample of the substrate on a glass or plastic plate and add a few drops of dilute hydrochloric acid or even vinegar. If the substrate material effervesces, it is proof that the material contains significant quantities of soluble carbonates. Such a substrate will eventually harden the water of any aquarium to which it is added and is not suitable for use with species that require soft, acid water to prosper.

In most instances, the chemical composition of the tank's substrate is a matter of indifference to riverine cichlids, while those native to Lakes Tanganyika and Malawi can only benefit from the use of materials rich in soluble carbonates on the tank base. Ordinary washed river or beach gravel suits the needs for the first group. Aquarists living in areas with extremely soft tap water often find it worth the extra expense of obtaining coral gravel as a substrate for Rift Lake cichlids tanks because of the extra buffering action if provides. Crushed oyster shell is not suitable for this purpose; it becomes too tightly packed, creating 'dead' pockets where anaerobic decomposition can occur. When disturbed, these can release toxic hydrogen sulphide into the aquarium, often with disagreeable consequences. A minimum layer of substrate should be used –

Below: *Since this substrate-spawning* Chalinochromis marlieri, *will dig at spawning time, be sure to plant* Anubias *in containers.*

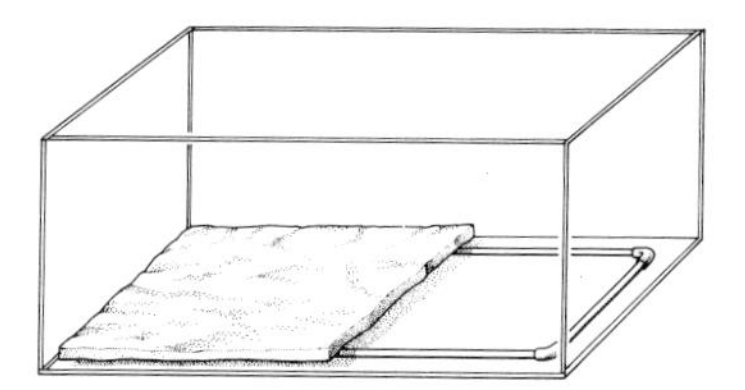

Above: *As a change from sand and gravel substrates, provide a piece of flagstone resting on a PVC pipe base. This novel form of tank decor is ideally suited to mbuna, the rockfish of Lake Malawi.*

enough to cover the base or filter plates.

As an alternative to traditional substrate materials, consider resting a single piece of flagstone, cut slightly smaller than the internal dimensions of the tank, on a framework of PVC pipes. This arrangement allows water to circulate freely under the slab, while offering the tank's less assertive inhabitants a refuge of last resort when harassed by more aggressive companions. It is particularly appropriate for an aquarium housing mbuna or their Tanganyikan counterparts, since most of these cichlids seldom encounter sand or gravel in nature.

Using live plants

The contemporary growth of interest in the culture of live aquatic plants is, on balance, a positive trend in aquarium keeping. Regrettably, it has been accompanied by a tendency to regard the unplanted aquarium as an 'unnatural' environment, in which fish are unlikely to prosper. The fresh waters of Africa are not richly endowed with aquatic plants, while biotopes dominated by them are very rare. The extreme seasonality of most savanna rivers imposes severe constraints on rooted aquatic plants. Only those species capable of surviving the dry season, such as tubers, bulbs or seeds, prosper under these conditions. Less hardy plants will thrive only in the relatively rare permanent marshes and oxbow lakes found in the lower courses.

Hardy plants suitable for cichlid tanks

Anubias barteri
Do no bury rhizome. Prefers moderate light. Herbivores find tough leaves distasteful. Hardy but slow growing.

A. nana
Dwarf anubias
Smaller than *A. barteri* but identical in requirements and growth pattern.

Aponogeton crispus
Wavy-edged swordplant
Provide bright light. Easily grown from bulbs. Tolerates a wide range of hardness and pH values.

A. ulvaceus
More likely than *A. crispus* to become dormant and shed leaves for a time. Otherwise differs only in leaf shape.

Bolbitis heudelottii
African fern
Do not bury rhizome. Prefers moving water and dimly lit tank. Distasteful to herbivores. Slow grower.

Ceratophyllum demersum
Hornwort
Provide bright light and thin out regularly.

Ceratopteris cornuta
African water sprite
Needs bright light. Easy to grow. Thin out frequently. Grows equally well floating or planted.

C. pteridioides
Water sprite; floating fern
As for *C. cornuta*. Does not thrive as a rooted plant.

C. thalictroides
Fine-leaved water sprite; Indian fern
As for *C. cornuta*. Does best planted, rather than floating.

Crinum thaianum
Thai onion plant
Plant neck of bulb above substrate. Leaves grow up to 1.5m (5ft) long.

Cryptocoryne wendtii
Prefers a moderate light. Tolerates hard, alkaline water.

C. willisii
Smaller than *C. wendti*, but also tolerates hard, alkaline water. Both species relish frequent water changes; avoid exposure to elevated nitrate levels.

Echinodorus major
Ruffled swordplant
Provide bright light; easy to grow in large (120 litre/30 gallon) tank.

E. parviflorus
Black Amazon swordplant
Easiest of the medium-sized swordplants to grow. All need bright light and tolerate high nitrate levels better than *Cryptocoryne* spp.

Hygrophila difformis
Water wisteria
Needs bright light. Pinch back emergent shoots to prevent loss of submerged foliage.

Microsorium pteropus
Java fern
Do not bury rhizome. Does best in dimly lit tanks. Distasteful to herbivores. Slow grower.

Nymphaea lotus
Red or tiger lotus
Needs bright light. Remove floating leaves to encourage vigorous growth of submerged leaves. Solid red, and maroon-speckled green leaved varieties also available.

Vesicularia dubyana
Java moss
Non-rooted. Prefers dim light.

Decorative materials suitable for cichlid tanks

Material	*Advantages*	*Disadvantages*
Driftwood Wide range of sizes available commercially. Do not collect pieces from local streams or seashore – they may contaminate the tank. *(C,S)*	Light, easy to work with; natural appearance. Carve out hollows to accelerate waterlogging and provide added shelter.	Buoyant; secure until waterlogged. May lower pH level of water.
Tropical woods eg mangrove. *(C,S)*	Denser than water so they drop immediately to bottom of tank. Suitable for West African forest-dwelling cichlids.	May discolour and acidify water. Soak for a week in 15 per cent solution of bleach, replaced daily, then rinse thoroughly. Not for Lake cichlid tank.
Basalt, tuffa *(A)*, **sandstone** *(C)*	Light; easy to work with	Not always available
Limestone. *(A,C,S)*	Eroded stone is lighter and provides colour. Calcareous. Provides additional buffering capacity.	Heavy; occupies substantial volume in tank.
Shale flagstone, slate, water-worn rounded boulders (agate, granite) *(A,C,S)*	Makes good caves and overhangs. Chemically neutral.	
Plastic and ceramic eg imitations of rockwork/driftwood. *(A,C,S)*	Readily accepted by fish. Easy to acquire. Do not look too artificial.	Providing a sufficient quantity can prove expensive.
Snail shells *(A)*	Calcareous; safe for Rift Lake cichlids. Many lamprologine species rely on them for shelter and spawning sites. Shells for marine tanks also suitable for cichlids.	Not suitable for soft water species.
Miscellaneous Clay flowerpots, empty coconut shells; sections of PVC pipe. *(A,C,S)*	Light and portable. Easy to modify for tank use.	Less decorative than natural materials.

Suitable for: (A) African, (C) Central American, (S) South American

Although forest streams do not exhibit such a seasonal pattern of flow, they are similarly lacking in rooted plants because their currents are often swift, their waters are poor in nutrients and often heavily shaded by the overhanging canopy.

Many cichlids will live happily in a well-planted aquarium, although a profusion of rooted plants is hardly essential to their well-being. Most species, however, appreciate a layer of floating plants in their tank. Such cover affords shy species a sense of security and goes a long way towards bringing them out into the open, where they are easier to observe. However, more elaborate attempts at aquascaping are hardly essential. Even forest-dwelling dwarf species, such as the several *Nanochromis* and *Pelvicachromis* species, will breed freely in a tank containing nothing more than a layer of water sprite (*Ceratopteris pteridioides*) at the surface and a few clumps of Java moss (*Vesicularia dubyana*) and Java fern (*Microsorium pteropus*) scattered attractively among the furnishings.

Another factor to consider when choosing live plants for a cichlid tank is the tendency of many species to treat such aquascaping as a self-service salad bar. All tilapias, for example, are herbivorous to a greater or lesser degree. Even species that do not normally feed on aquatic plants in nature will browse on them in captivity. With the exception of the several *Labidochromis* species and the long-snouted, predatory *Melanochromis* species, so will the colourful mbuna of Lake Malawi and their Tanganyikan counterparts of the genera *Petrochromis, Simochromis* and *Tropheus*.

Planting

When it comes to discussing plants and large cichlids, most fishkeepers throw up their hands in despair; the two are often thought to be incompatible. Certainly, dainty plants are very quickly devoured by large cichlids on the rampage or in search of food. Nonetheless, some species of plants are inedible or sturdy enough to survive the rigours of a cichlid system.

The evergreen Java Fern (*Microsorium pteropus*), *Hygrophila corymbosa*, the False Cryptocoryne (*Spathiphyllum wallisii*) and the Amazon Sword (*Echinodorus* sp.) are all tough. Java Fern is a particularly hardy plant. It will thrive in alkaline, even brackish, water and would be difficult for even the largest cichlids to consume.

Combining suitable real and

Below: *This fast-flowing highland river in Costa Rica illustrates the rock rubble habitat that an ideal aquarium should simulate.*

Above: *This aquarium uses plastic plants and rounded boulders to good effect. Combining real and plastic plants also looks natural.*

artificial plants can also be an extremely effective strategy. The art of using artificial plants successfully in a living aquarium relies on creating the impression that the plants are growing. Grouping the plants in clumps produces a convincingly natural effect. A 30-38cm (12-15in) plastic *Vallisneria*, for example, looks more 'alive' with two or three 15cm (6in) versions arranged in a cluster around the base. After a few weeks, algae forms on the synthetic fronds and the illusion is complete.

Layouts for small cichlids

Based on an aquarium measuring 60 × 38 × 30cm (24 × 15 × 12in) or 90 × 45 × 30cm (36 × 18 ×12in), the layout on pages 52-53 is ideal for keeping small cichlids. It would be an eminently suitable layout for fishkeepers new to cichlids. Use a basic substrate of rounded gravel with a particle size of approximately 5mm (3⁄16in) and to a minimum depth of 7.5cm (3in).

The main 'architectural' feature of the layout consists of a cluster of rocks and slate. Build this up at one end of the aquarium using water-worn rocks from hill streams or hard, sea-washed (and suitably rinsed) rocks from the beach. It is important not to make the cluster too high because spawning cichlids will dig out gravel at the base and may cause the stones to topple and fall against the aquarium glass. Prevent cichlids digging too far down into the gravel by placing a sheet of gravel tidy about 7.5cm (3in) below the surface of the gravel. Nevertheless, ensure against the possibility of an underwater rock fall by looking for a single piece of well-jointed rockwork to form the centrepiece of the arrangement. Finally, butt up pieces of slate to the rounded boulders to create a series of 'natural' crevices the fish can use.

Litter a few pebbles in the centre of the aquarium floor for the fish to 'associate' around and decorate this area with a real Sword Plant (*Echinodorus* sp.) or its plastic equivalent. Arrange some pebbles at the opposite end of the tank, perhaps more than the token amount used in the centre, and frame an upturned or side-down plant pot with living or artificial *Vallisneria* plants. A small piece of bogwood completes the scene.

In a 60 × 38 × 30cm (24 × 15 × 12in) aquarium you will need only one undergravel filter plate and its associated air-operated uplift. For a 90 × 45 × 30cm (36 × 18 × 12in) tank, use two equally sized filter plates; position the uplifts at opposite ends to enhance the flow of water around the aquarium. Use power heads at the top of the uplifts, position these side by side in the centre back of the aquarium and conceal the whole arrangement with a single piece of bogwood. The powerful flow of water produced by the power heads will be more than sufficient to distribute the heat from the heater-thermostat.

Layouts and aquascapes for larger cichlids

Large layouts (as that on pages 54-55) allow more swimming space for larger cichlids. Suitable aquaria measure 120 × 45 × 30cm (48 × 18 × 12in) or 120 × 45 × 38cm (48 × 18 × 15in). The focal point should be an arrangement of rounded boulders framed by a large overhanging piece of bogwood. If a suitable single piece of bogwood cannot conveniently be found, build up the desired shape by gluing several smaller fragments together with aquarium sealant.

To one side of this structure build up a 'cavework' with small boulders and a roof of slate. This will provide territory for a single pair of cichlids. If a group of juvenile cichlids are continually fighting to move up the pecking order or to establish territory, it is advisable to make up an extra cluster of rockwork at the other end of the tank. In most cases, it is more preferable to provide a surplus of shelter/territorial options rather than just a sufficient number. The open areas left in the aquarium will allow the cichlids to spawn, perform and generally posture against each other as they do in the wild.

Use a gravel substrate as specified in the first aquascape. Fit two undergravel filter plates in the aquarium and operate them with a simple air lift and power head respectively. This will provide a beneficial combination of water flow and aeration around the aquarium.

Above: *An aquascape based on a 60×38×30cm (24×15×12in) tank.*

Right: *A spawning pair of* Cichlasoma spilurum *the larger fish is the male. This species is well suited to medium-sized aquariums, and such a pair could successfully raise fry in the aquascaped tank shown above.*

Discus prefer deep, still water aquaria with vertical shelter in the form of wood spindles or bogwood. To provide the correct conditions it is best to select a lengthy aquarium with an above average depth and front to back measurement of about 50cm (20in). With such an aquarium it is

Rounded boulders and a piece of slate create a natural series of crevices for the fishes to use.
A half-buried pot provides an ideal refuge or breeding site for territorial fishes.
Bogwood – an essential part of any Central American system.

possible to provide filtration that can be concentrated to one side of the system allowing a 'quiet water' and a light-reduced 'shaded' area to the other.

Larger aquascapes for the largest cichlids

A few specimens of the larger species, such as *Cichlasoma motaguense* or *C.managuense*, or a group of the smaller species would find the expansive aquascape illustrated on pages 56-57 a true 'home from home'. The layout is based on an aquarium measuring 150 × 45 × 38cm (60 × 18 × 15in). At first sight it seems cluttered. The combination of boulders, bogwood

and general litter, however, creates an environment in which the strong can form territories and the weak can escape the attentions of bullies.

Use three undergravel filter plates in the base of the aquarium and cover these with at least 7.5cm (3in) of rounded gravel as recommended for the other aquascapes. Place a large piece of bogwood in one corner of the aquarium and arrange a beechwood branch so that it extends into the centre foreground. Along the back of the tank build up a wall of large boulders, taking care to make sure they are stable. In the opposite corner, arrange rounded boulders

Below: *A substantial piece of bogwood lies at the heart of this aquascape for a 120cm (48in) tank. The result is spacious but provides plenty of territorial refugees.*

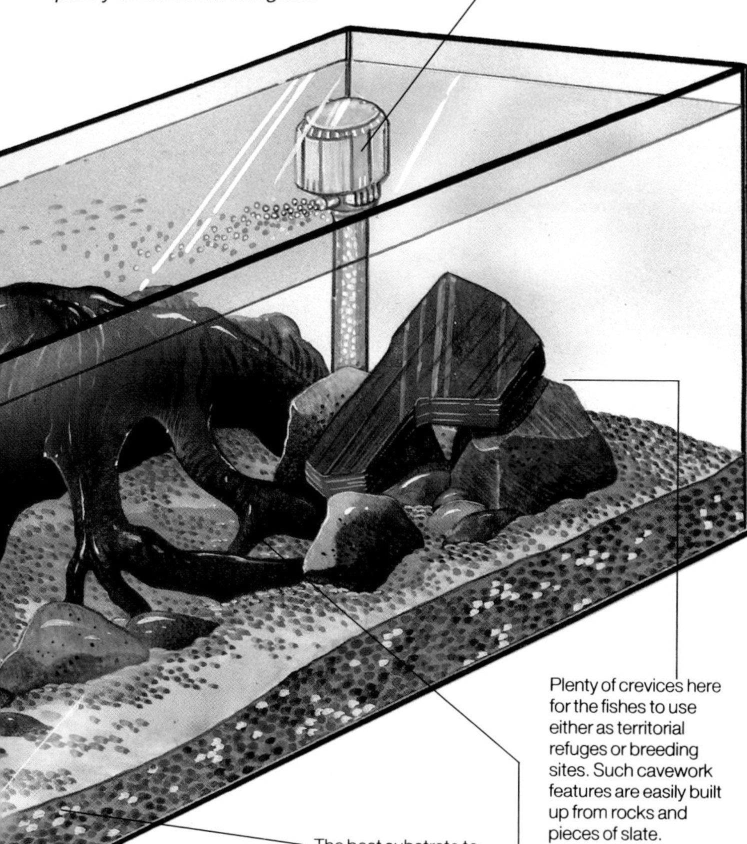

and pieces of slate to create crevices, caves and holes. Position plastic plants (Sword Plants are ideal) here and there to complete the scene.

Using a combination of air-operated and power head uplifts for the undergravel filters satisfies the twin demands of an aeration and good water flow necessary in such a large aquarium. If you direct the outlet of the power head located behind the cave area across the back of the aquarium towards the airlift tube in the centre, this will drive air bubbles around the aquarium and create a water flow towards the power head in the opposite corner. This second power head will move water towards the front glass and set up a beneficial circulation around the entire aquarium. Fish will often swim around, along or into the powerful streams of water issuing from the electrically driven power heads.

Since cichlids tend to be

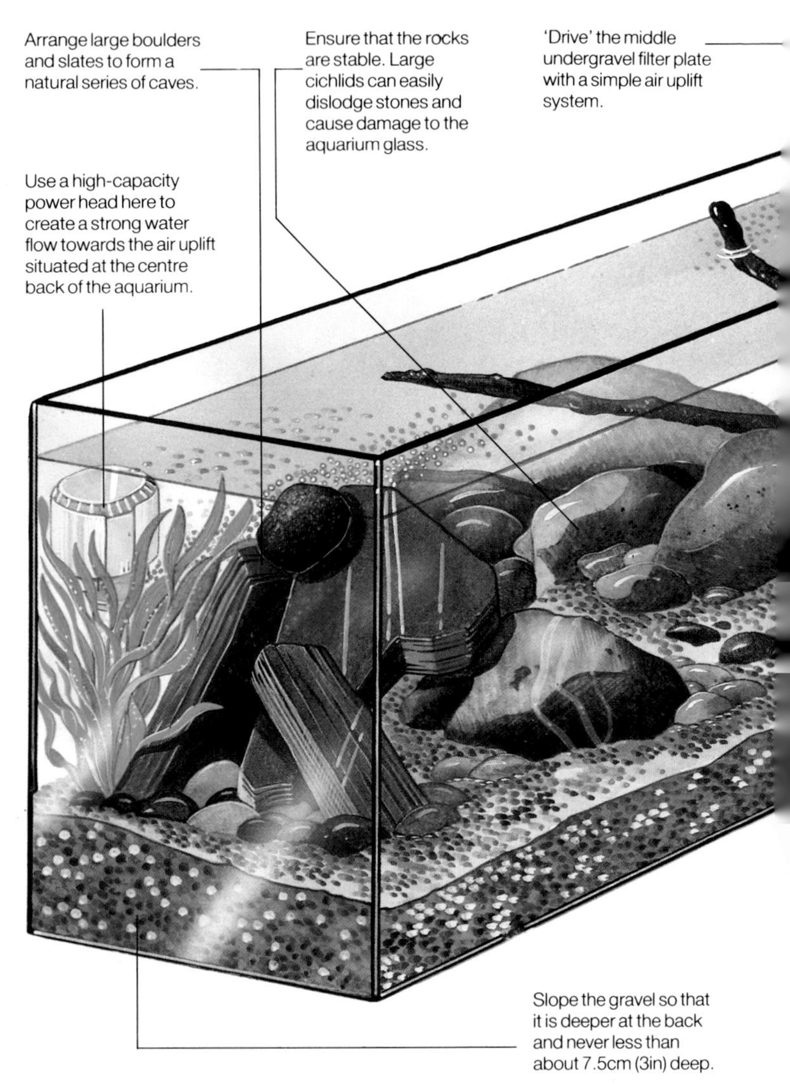

aggressive and sometimes downright belligerent, it is appropriate to provide shelter where harassed individuals can escape. This can be a simple tangle of driftwood, a pile of rocks, broken or inverted clay flowerpots, plastic PVC piping of appropriate diameter and length, or any number of innovative and non-toxic solutions. Keep watch for signs of battering and be prepared to remove or separate combatants. In the wild, beaten fish simple swim away. In the confines of the aquarium, they are often killed.

Aquascapes for dwarf cichlids
Dwarf cichlids can be housed in smaller tanks. A compatible pair of most species will get along in a 45-litre (10-gallon) tank provided there is adequate shelter. However, as is true for most situations, larger tanks are better for reasons of

A beechwood branch and large piece of bogwood create a focal point in the design. Behind the bogwood use a power head to direct water to the front.

Above: A large varied aquascape for 150cm (60in) aquariums, allowing plenty of swimming room.

Right: An adult *Cichlasoma managuense*. Large specimens need space and can readily dominate their tanks.

water stability. Long 90-135 litre (20-30 gallon) aquaria are recommended. Filtration should be via a combination of sponge, undergravel and/or canister with additional aeration essential because of the higher maintenance temperatures of 25-30°C (78-86°F). Many dwarf cichlids are harem polygynists – one dominant male spawning with several females – with each female holding a separate territory. Thus, adequate shelter in the form of flowerpots, or other caves is recommended. A well-planted tank, including rooted and floating surface plants to cut down on top illumination, is recommended. Dwarf cichlids rarely dig. They may be kept with other peaceful 'community tank' fish like tetras, hatchetfish, pencilfish and bottom scavengers like *Corydoras* sp. catfish or small loracariids (eg *Peckoltia* sp.) The use of such dither fish helps make these retiring cichlids less shy. Attention to water chemistry (soft, acid) and water quality is essential for success with these fish.

Mouthbrooding and medium-sized Acaras

With the exception of slightly larger tanks, mouthbrooding Acaras of the genus *Bujurquina* prosper under the set-up recommended for dwarf cichlids.

Discus

The aquascape that is suitable for the 'Kings of the Amazon' would also suit similar shaped fishes and many dwarf cichlids. Often it is an above average water depth that is emphasised for discus, however, as territorial fishes, they also thrive when offered plenty of substrate space. To help maintain good water quality, external power filters are recommended and, as fussy feeders, they are best kept on a shallow substrate without undergravel filters. Uneaten food will sit on top of sand or fine silica gravel whereas it is drawn into the undergravel filter. This is not a hard and fast rule as many discus keepers have successfully employed an undergravel filter in their systems. It is widely agreed that undergravel filters require regular syphoning to extract sediment and this activity is not always appreciated by nervous cichlids. Discus appreciate a simple aquascape combined with subdued lighting and vertical pieces of wood branches in which they can take cover.

The same deep water system would also suit large Altum angelfishes. Urau and Severums, the community could also include medium sized, peaceful charcins (large cardinals, hatchet fishes, sparkling tetras, bleeding hearts etc) and small to medium catfishes (*Corydoras, Brochis, Dianema* and loricariids such as whiptails, *Peckoltia, Anestrus* etc).

Aquaria for Eartheaters

Eartheaters sift the substrate for edible detritus, and have some special needs as aquarium fish. They should be kept over sand or fine gravel so they can practise what they were designed to do. However, this habit makes rooted plants an impracticality. Both Java Fern (*Microsorium pteropus*) and Java Moss (*Vesicularia dubyana*) are recommended as is some form of floating plant to cut down on surface light intensity would be ideal. Shelter in the form of a tangle of bogwood which also provides a site of attachment for the plants works well and leaches the tannic and humic acids which are welcomed by these fish. Again, tanks with larger bottom areas are preferred to 'high' tanks of a similar capacity. Water should be on the soft and acid side and kept very clean and relatively warm: 25-29°C (78-83°F). The fish are fed a variety of prepared, pelleted (sinking) foods as well as frozen foods (bloodworms) and occasionally live foods, particularly worms. Larger dither fish in the form of silver dollars (*Mytennis, Myleus* species) or elongated hatchetfish (*Triportheus* species) are recommended as are most catfish.

Above: *Characins, like this Diamond Tetra* (Moenkhausia pitteri), *make excellent dither fish for dwarf and medium-sized South American cichlids.*

Pike cichlids

Contrary to reputation, many pike cichlids are reasonable fish for the cichlid community. They can be aggressive, but usually only with members of their own species. To this end, communities containing a number of pike cichlids should restrict the population to pikes of near identical size, and should provide adequate shelter in the form of PVC piping cut to size. As a group, the majority of pikes are undemanding when it comes to water chemistry or even quality. However, they are voracious and messy feeders, and good mechanical filtration, in the form of external power filters, is recommended. Live feeder fish are necessary for only the most stubborn of wild-caught specimens: most pikes can be converted to floating freeze-dried krill, frozen bloodworms, even prepared foods.

Tankmates, should be limited to fish that are large enough so that they cannot be swallowed: a recommendation can also be given to other cichlids of similar temperament and even large silver dollars.

Other large Neotropical cichlids

Many of the large species, typically 'cichlasomines', pose problems in maintenance, with the two main problems being managing aggression and maintaining water quality. There are at least two approaches to managing aggression in the large cichlid community. One is to keep them relatively crowded in tanks devoid of any shelter which diffuses the aggression amongst a large number of individuals and which provides no visual markers (rocks, driftwood, etc) for establishing territory. The result is often a surprisingly peaceful group of thugs. The second approach is to maintain several pairs of different species with adequate shelter (overturned flowerpots, etc) and tank space and let the pairs establish territories. With other pairs serving as targets, pair bonds are enhanced. If enough room is provided, most of the aggression will take the form of ritualized display and threat with little harm done to the combatants. Such an arrangement often promotes spawning of otherwise highly aggressive, unspawnable fish. Sometimes individual rogue fish must be maintained by themselves.

While the actual details of water chemistry and quality vary from species to species (see the South

American cichlid species catalogue), one general point should be re-emphasised: big cichlids make big messes! As such, mechanical filtration in the form of several external power filters and regular, large water changes are essential for the successful maintenance of the larger species. Typically these fish are omnivorous and will eat a wide variety of prepared pelleted, freeze-dried and frozen foods. Earthworms are a great treat.

Tankmates for African cichlids

With the exception of certain predators, such as the banded jewel fishes, *Cyphotilapia frontosa*, the larger *Lamprologus* species and a number of haplochromine cichlids native to the Great Lakes, sexually quiescent African cichlids will usually ignore non-cichlid tankmates that are a third to one half their overall length, or even larger. Since cichlids prefer the lower third of the water column, midwater-swimming fishes fill a visual gap in a tank. The untroubled behaviour of schooling species also adds greatly to shy cichlids' sense of security; indeed, the judicious addition of such dither fish to their aquarium is a well established technique for bringing the most retiring dwarf cichlids out into the open.

Suitable dither fish must be too large to tempt the appetites of their cichlid neighbours, yet not so large that they offer serious competition at feeding time or pose an unreasonable threat to cichlid fry. Small to medium-sized danios, barbs or rasboras make good dither fish for forest-associated cichlids, since they prefer the same water chemistry. The smaller labyrinth fishes also make good tankmates for West African dwarf cichlids. Poeciliids and Australasian rainbowfishes are at home in the hard, alkaline water demanded by Lake Tanganyikan cichlids and make suitable companions for all but the most predatory cichlids. Goodeids also find such water conditions to their liking, but be cautious when housing them with small cichlids. Some of the more robust goodeids are so aggressive that they function as an anti-dither fish, terrorizing their cichlid neighbours so badly that they refuse to leave their hiding places any longer than is necessary.

Few traditional scavenger fishes make good tankmates for African cichlids. Loaches of the genera *Botia* and Noemachelius and the 'sharks' of the genus *Labeo* have the same requirements for shelter as small and medium-sized cichlids, and generally assert their rights forcefully, often with disagreeable results for the cichlids. Corydoras and other mailed catfishes are a target for cichlid harassment and do not do well, even alongside the smaller

Below: *Pike cichlids, such as this* Crenichla *species captured in Guyana, can be kept providing there is shelter and small species are not included.*

African cichlids. This vulnerability extends to most small smooth-skinned catfishes, although in nature their larger representatives, in turn, often become significant cichlid predators!

The two catfish groups most likely to prosper in an African cichlid tank are the upside-down catfishes (Family Mochokidae) and the armoured suckermouth catfishes (Family Loricariidae).

The most commonly available upside-down catfishes are *Synodontis* species, although representatives of the genera *Brachysynodontis, Hemisynodontis* and *Mochokiella* are sometimes found in importers' tanks. Upside-down catfishes also require access to shelter, but seem better able to coexist with African cichlids than many other territorial bottom-dwellers. They will clean up uneaten food from the bottom and, if hungry enough, many species will graze on algae as well. However, their chief role in a cichlid tank is to pose a potential threat to eggs or fry. In so doing, they provide a focus for the aggression that a breeding pair would otherwise direct towards one another. Obviously, you cannot choose upside-down catfish for the role of 'target fish' if they are much larger than the parental cichlids, otherwise they may overwhelm the cichlid defences.

Above: *Eartheaters, like this group of* Satanoperca, *thrive in aquaria that offer a sand substrate in which they can sift for food.*

Loricariid catfishes make ideal scavengers in an African cichlid tank. They can be counted on to clean up uneaten food promptly and, if kept a bit hungry, will do a good job of controlling algal growth on solid surfaces. As with upside-down catfishes, it is a good idea to select a loricariid of a similar size to the breeding cichlids in the tank. Slender-bodied species (*Farlowella* and *Rineloricaria* spp.) and such dwarfs as *Otocinclus* and *Parotocinclus* are too easily damaged to be risked with even the smallest African cichlids. In any event, the behaviour of both mochokid and loricariid catfishes varies considerable and it is always a sensible precaution to consult an appropriate reference before adding a given species to an African cichlid tank.

Maintenance

Routine maintenance in the cichlid tank includes cleaning the front glass, maintaining the filter, removing waste particles from the substrate and making partial water changes. The frequency with which these tasks are carried out depends primarily upon a tank's population density. Apart from the regular chores, monitoring the water (PH, nitrite levels, temperature, etc) and the fishes' behaviour is also paramount.

Above: *Lake Malawi cichlids are extremely sensitive to nitrogen cycle mismanagement. Efficient biological filters, regularly cleaned and serviced, are essential if they are to thrive in captivity.*

Algae

Algae on rocks can be considered natural and a beneficial influence in balancing the aquarium system. Algal cells absorb nitrate from the water as a food source (hence nitrate fertilizers) and provide a large surface area to support the oxygen-fed bacteria that 'drive' the nitrifying cycle. Algal growths can also form a useful food source for certain cichlids; *Neetroplus nematopus* is one of the cichlids that has evolved teeth structures to enable it to browse on algae.

Rampant algae, however – in some varieties filamentous algae – can be unattractive and harmful to the system. An excess of algae usually shows that nitrates are high and the lighting is too strong or has been left on too long. Direct sunlight striking the tank at certain times of the year may also cause algae problems. Sunlight creates a pleasing effect and should not be avoided completely, but if it causes unsightly 'algal blooms' across the substrate and glass, then you must take steps to control the amount reaching the tank.

Below: *A magnetized scrubbing pad is a convenient and effective means of removing algae growth from the front glass of the tank.*

Every three or four days, clean the inside surface of the front panel, using a magnetized scrubbing pad. Many algae species are easy to scrub away when newly settled onto a solid surface, but become progressively less easy to remove once established. Stubborn algae requires the attentions of a razor blade, a much messier and less agreeable process! A good growth of green algae on the back and sides of a tank plays a useful role in nitrogen cycle management and provides herbivorous cichlids with a valuable dietary supplement, so do not be too quick to remove it. Uncontrolled growth of blue-green algae results when a tank is overstocked and receives too much light. The only way of controlling it is by tackling the root causes of the problem.

Water changes and removing waste
African cichlids differ considerably in their tolerance of water changes and it is important to bear this in mind when planning and carrying out such a programme. Do remember that the more heavily stocked an aquarium, the more frequently it will require water changes, regardless of the species it houses. It is a simple matter to combine a water change with cleaning the tank bottom. Unless a tank houses large, very messy cichlids, it is not necessary to vacuum the entire bottom with each water change. As long as the substrate is turned over once a month, serious problems are unlikely to develop.

Maintaining filtration systems
Every week check the power head or power filter flow for strength. A reduced flow on a power head signals a blockage in the undergravel filtration system and the time may have arrived to clean the gravel. Use one of the inexpensive syphonic action devices to draw up debris from between the gravel particles. Since the organic debris is lighter than the gravel, it is whisked away in the syphonic flow of water set up in the funnel while the gravel remains in the aquarium. Cleaning the gravel in this way every three or four weeks will prevent accumulating dirt from disturbing the pH balance of the water and ensure the free passage of oxygen so essential for the nitrifying bacteria to flourish. It is the equivalent of a gardener turning over the soil.

Cleaning power filters is also a regular but less frequent task. Leave power filters running until the flow rate is clearly reduced by 60-70 percent. This indicates a significant build up of sediment inside the canister. To clean the filter, empty all the contents into a bucket of water taken from the aquarium, rinse the medium and filter wool thoroughly and return them to the canister. The important point here is the use of aquarium water to clean the filter contents. Avoid the temptation to wash the filter medium under running tap water; the chlorine in the water can suppress the bacterial activity established in the filter. Rinsing the medium in aquarium water of the correct pH and temperature does not disturb this bacterial activity.

Inspect airstones used in undergravel filter uplifts. A noticeable reduction in aeration usually points to the deterioration of an airstone. These are cheap to replace and if left on the system when blocked they can reduce the efficiency of both filtration and pump. A blocked airstone can put back pressure on the air pump, for example, and this in turn will shorten the life of valves and diaphragms.

Checklist

The following checklist will simplify the task of monitoring a cichlid tank:

- Are all the filters and airstones functioning normally? If not, why?
- Is the water temperature within the acceptable range for the tank's inhabitants?
- Is the tank's water clear, with a fresh smell?
- Is the number of fish visible equal to the number supposedly present in the tank? If not, why?
- Do the fish appear different in any way? Has their colour pattern changed? Is their rate of respiration elevated?
- Is the fish's behaviour different in any way? Do they hang motionless below the water surface? Do they dash frantically about the tank at the slightest disturbance and try to hide? Do they refuse to eat when fed? Have the fish become more aggressive towards each other?

The answers to several of these questions are self-evident. Always restore filter or pump function immediately; adjust thermostat settings up or down; locate missing specimens; remove dead fish from the aquarium and prevent obvious bullies from abusing their tankmates. Other signs require further investigation before the underlying problem – and appropriate solutions – can be identified.

Monitoring water

Cloudy, greyish water, often associated with an unpleasant smell, indicates biological filter failure. It is often accompanied by a change in the appearance and behaviour of the fish, such as an exaggerated rate of respiration, darker than normal coloration and 'panic' swimming in response to the slightest disturbance. 'Grey

Below: *Regularly removing organic detritus from the substrate is simple using a siphonic gravel cleaning device such as this.*

Below: *Always treat 'raw' tapwater with a good-quality conditioner to reduce levels of chlorine and other 'disinfectants' in the water.*

water' usually appears immediately after a tank has been set up. It is caused by adding a full complement of fish to the aquarium before the filtration system is fully established. However, even a totally mature biological filter bed can 'crash' if its waste handling capacity is exceeded through overfeeding or by adding 'just one more' fish to the aquarium.

In the short term, address these symptoms immediately by removing any obvious decomposing organic matter – such as uneaten food or a dead fish – from the aquarium, making a 60-75 percent water change, thoroughly cleaning the substrate and cleaning the filter(s). It is also wise to add a chemically active medium to outside filters before refilling the tank with fresh water. In the long term there are two solutions to this problem: you can either reduce the waste load by cutting back on the number of fish stocked, or increase the filter's waste processing capacity by adding another filter unit to the system.

Aquarium water that remains clear, but assumes a yellowish or amber tinge may also indicate nitrogen cycle management problems. The first response to such a development should be to check the pH and nitrate values of the water. An acidic reaction – especially in a sample taken from an aquarium where neutral to basic conditions should prevail – clearly indicates that something is amiss. Unless a new piece of driftwood has recently been added to the tank, the most likely cause of acidification is a build-up of metabolic waste. If a nitrite test yields readings of more than 0.01ppm assume that metabolite accumulation is the culprit.

Gradual acidification is most likely to occur in areas with soft tap water. Apart from the possibility of nitrite poisoning, a drop in pH level clearly poses a more serious threat to the well-being of Rift Lake cichlids than to West African species. If the aquarium contains an alkaline substrate, such as coral gravel, or added crushed coral or oyster shell in the filter and regular water changes are made, then pH drops should be very rare. More frequent water changes, combined with scrupulous attention to filter maintenance, will prevent these symptoms in the short run. In the long run, only reducing the number of fish housed in the tank or increasing filter capacity will resolve this problem.

Large-scale water changes tend to shift the pH into the alkaline range. Such a shift converts harmless ammonium ions into toxic ammonia. To minimize the risk to the fish of ammonia poisoning, either add a chemically active medium to the filter system before adding fresh water to the aquarium, or use a one-step dechloraminating agent in conjunction with the water change.

Signs of acute respiratory distress, such as 'gulping' at the water surface, and marked colour pattern changes sometimes occur without an obvious cause, ie such as a pump or filter malfunction or loss of water clarity. In such a case it is worth investigating either biological filter failure or contamination of the water by some external agent, such as insecticides, household cleaning products or paint fumes. Before taking any corrective action, test the nitrite and ammonia levels. A positive test result (ie a concentration of more than 0.01ppm) indicates that excessive metabolite concentration is the probable cause of the problem and you can then implement corrective measures.

A negative test, however, indicates that an external agent is at work. In this event, immediately remove and discard any activated carbon in the filter unit, add a chemically active medium to the system and carry out an 80-90 percent water change. If the fish can be saved, this should alleviate their obvious symptoms and give you sufficient time to locate and remove the source of the problem. Many insecticides are toxic in minute concentrations, so it may prove necessary to repeat the water change several times to eliminate all traces of them from the system. Once the symptoms have disappeared, remove the chemically active medium from the system and discard it.

Checking pH and temperature
Take pH and temperature readings every week. A falling pH (ie tending towards acidic conditions) usually indicates a need to extract sediment from the substrate and power filter media as described above. Normally, regular water changes combined with the removal of sediment will prevent serious changes in the pH value of the water. Overstocking and excessive feeding of large fishes will also cause a falling pH value.

Water temperature can range widely if the room temperatures are extreme. In normal circumstances, large temperature fluctuations point to the failure or inefficient operation of the heater-thermostat or suggest that its power rating is inappropriate for the intended use. To guard against excessive cooling, some fishkeepers install a back-up heater-thermostat set to a lower temperature than the main unit or units. The majority of temperature regulation problems seem to cause the opposite effect, however. Experience shows that the most common fault is the failure of the thermostat to cut out the heating element. The subsequent rise in temperature usually reduces the oxygen levels in the water to unacceptable levels and fish are lost as a result. Get into the habit of placing a hand on the aquarium glass to gauge the relative warmth of the water, especially last thing at night or early in the morning. If the tank feels warm relative to a normal body temperature of approximately 37°C (98.6°F) then check the thermometer immediately.

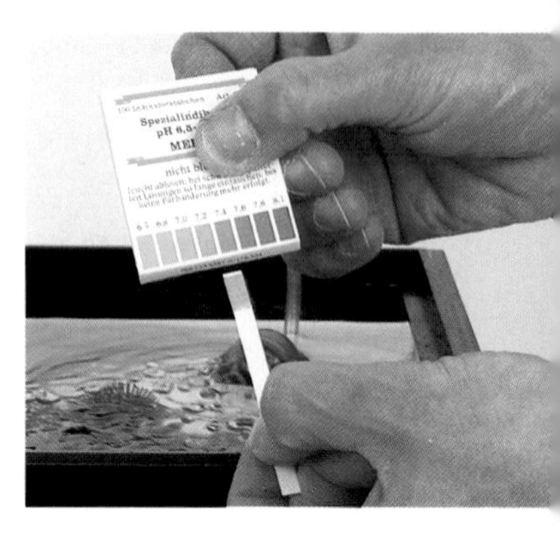

Above: *This paper strip indicator provides a quick way of taking a pH value reading. Monitor tapwater and aquarium water pH values regularly.*

Testing nitrite/nitrate levels
It is advisable to test for nitrite levels during the first six weeks of an aquarium's life – while it is still maturing in the bacterial sense – or if you suspect that the filtration system has broken down. Such breakdown might show itself in terms of cloudy water and/or the fish showing signs of itching. If an excessive nitrite peak occurs during the maturing process (a transient peak is natural), try introducing gravel from a mature system or use a proprietary additive designed to enhance the nitrifying cycle.

In the established cichlid aquarium it is essential to test nitrate levels on a weekly basis. High nitrate readings (above 25ppm) indicate stocking levels are at their maximum for the size of aquaria and that larger than normal partial water changes are necessary to reduce the build up of organic wastes in the system.

Some regions in the UK record high (up to 50ppm) nitrate levels in the domestic tap water supply. In this situation it is advisable to use a resin exchange cartridge 'filter' or add specific NO_3 removing resin to the filter. Fresh water should be used during water changes to reduce nitrate levels at source. Many African cichlids appear to be susceptible to high nitrates and

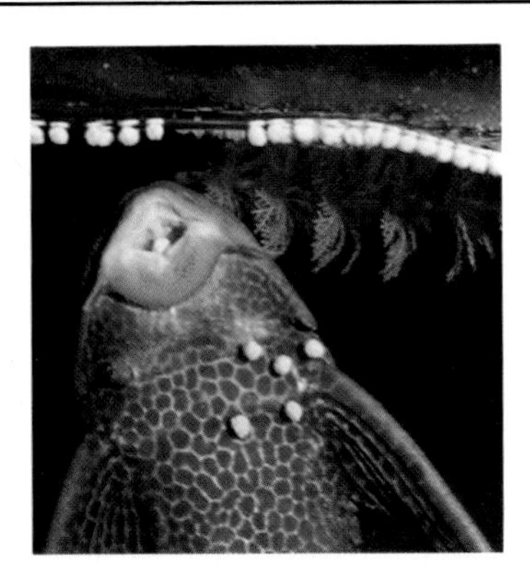

Above: *A large Suckermouth Catfish (*Pterygoplichthys *sp.) takes excess cichlid pellets from the surface. An ideal scavenger for a cichlid tank.*

such water imbalances may contribute to the Malawi and Eartheater bloats referred to in cichlid literature.

If nitrite or NH_3 levels remain unnaturally high in an established system, discuss the matter with your dealer, who should be able to offer valuable insights into the problem related to your particular set-up.

Help with the 'housework'

Large cichlids produce copious amounts of waste and are inefficient feeders, especially of prepared foods such as pellets or foodsticks. When eating these foods, cichlids spit or blow out clouds of food particles. This increases pressure on the 'cleaning' processes (natural or assisted) that take place within the aquarium and highlights the need to monitor water chemistry very closely and carry out regular, partial water changes.

To help with the 'housework' in a community system containing large South American cichlids, introduce a catfish as a substrate scavenger. Members of the armoured Catfish Family, Loricariidae, are tough enough to withstand the close aggressive attentions of cichlids. They will eat almost any type of food, including algae, but will not predate on fish eggs. The Suckermouth Catfish of the genus *Hypostomus* (of which *H. plecostomus* is one of over 100 species known) are ideal catfish to include in a community of large cichlids. They use their rasping teeth to scrape algae from rocks and other surfaces.

Scaleless catfish of the genera *Rhamdia* and *Pimelodella* – both found throughout South and Central America – are also suitable for alkaline water, large cichlid community systems. Unfortunately, they *will* eat eggs and fry, presuming the brooding parents are caught off guard long enough. If spawning and rearing cichlids are priorities, then do not include *Rhamdia* or *Pimelodella* in the aquarium.

Rift Lake cichlid systems often include members of the *Synodontis* genus relevant to the particular lake. They are excellent 'scavengers', in the true sense of the description (they will not eat rubbish!), but problems can result in certain communities. Breeding cichlids will find them nothing but a nuisance and fry may be lost as the catfishes hunt for food during nocturnal activity.

Riverine cichlids will thrive amongst the many non-lake species of *Synodontis* although it is often better to keep with small species as many grow quite large and can become boisterous in a small fish community.

Overstocked, under-sized aquaria will always present problems. If you use a large enough tank with sufficiently powerful 'life-support' systems, regular maintenance will keep cichlid communities in good condition.

It must be stressed that there must be a strict regime of regular partial water changes if the community is to thrive. Despite the best filters, aged water will deteriorate rapidly if fish growth and eventual stocking levels are not allowed for in the initial calculations. The greatest myth in fishkeeping is that healthy aquaria look after themselves.

Feeding

Cichlids fall into one of four subgroups in terms of their dietary requirements: omnivores, micropredators, piscivores and herbivores. It is important to satisfy the particular nutritional requirements of each group if the fishes are to thrive in captivity.

All cichlids tend to gluttony in captivity, although such behaviour is seldom a reflection of their nutritional state. In nature, many species rely significantly on high-roughage foods, such as algae or organic detritus. Although often abundant, such resources supply little food value per gram and in order to meet its metabolic needs the fish must eat continuously. At the other extreme, large predators exploit relatively scarce prey items of high nutritional value. In the wild they seldom, if ever, have the opportunity to satiety, so natural selection has never operated to place a limit on their feeding behaviour. Hence, these fish will also overeat whenever afforded the chance.

It is obviously essential to offer any captive animal sufficient food to enable it to grow normally, remain healthy and, if the opportunity arises, to breed successfully. However, you need not feed to excess to accomplish these ends. Overfeeding high-protein, low-roughage foods may actually harm a cichlids's health and will inevitably make nitrogen cycle management in the aquarium more difficult.

The high quality and range of prepared and frozen foods now available means that fish in captivity are probably in a better condition than those fending for themselves in the wild. Also, if water quality is maintained many, if not all, fish live longer in aquaria than in their natural habitat. In the aquarium, they are protected from droughts and floods and have food placed before them every day! In the wild, food is not always constantly available; because of season or circumstance some days it may be plentiful; some weeks it may be scarce. To echo this natural irregularity, consider leaving the fish unfed on one day each week. Although this strategy is open to debate, there are sound reasons behind it. In most cases, aquarium fish are overfed. Thus, missing a day tends to take the pressure off the system in terms of dealing with organic waste. If you have spawning or brooding fish in your aquarium, however, do not interrupt the feeding routine. Parent fish take in food and blow out any excess through the gills; this is an important source of food for fry.

In their natural habitat, most non-specialized feeders cichlids eat plant debris, young fish, crustaceans, insect larvae, and terrestrial insects and worms. In the aquarium, these fish will readily accept prepared foods, such as large flake foods, pellets and foodsticks, as the bulk of their diet. Feeding a selection of frozen and live foods on a rota basis, such as brine shrimp, *Mysis* shrimp, earthworms and bloodworms, will keep the fishes in excellent health.

Variety is the key factor for successful feeding. Offer established fish communities several feeds during the daytime or evening and alternate the type of food. On any one day, for example,

Below: *Even herbivorous cichlids require a weekly feeding of animal protein. Here* Tropheus *sp. relish an offering of red mosquito larvae.*

Feeding cichlids

Feeding group	*Recommended diet*
Omnivores *Anomalochromis thomasi, Chaytoria joka, Chromidotilapia* spp., most small *Haplochromis* spp., *Melanochromis johanni, Nanochromis parilius, Oreochromis* spp., *Pseudocrenilabrus multicolor, Thysia ansorgii, Tilapia* spp. Many dwarf or small South American cichlids	High-quality flaked/pelleted foods; proprietary conditioning foods with higher concentration of vegetable matter; frozen bloodworms; glassworms; zooplankton; seasonally available livefoods, eg adult brineshrimp (*Artemia*) and water fleas (*Daphnia*); mosquito larvae. Two feeds daily, or more frequent smaller offerings.
Micropredators *Aulonacara* spp., *Chilotilapia rhoadesi, Chromidotilapia batesi, Cyathopharynx furcifer, Cynotilapia* spp., *Cyprichromis* spp., most small to medium-sized *Malawi Haplochromis* spp., the small *Hemichromis* spp., *Iodotropheus sprengerae, Julidochromis* spp., *Labidochromis* spp., small to medium-sized *Lamprologus* and *Telmatochromis* spp., *Nanochromis* spp., *Pelvicachromis* spp., *Tanganicodus, Spathodus* and *Eretmodus* spp., *Steatocranus gibbiceps, Teleogramma* spp., *Xenotilapia* and *Callochromis* spp.	High-protein flakes or pellets; freeze-dried or frozen zooplankton; frozen bloodworms; chopped earthworms; live *Daphnia, Artemia*, glassworms, mosquito larvae and snails. Three light feeds daily. Live foods encourage breeding.
Piscivores *Boulengerochromis microlepis, Cyphotilapia frontosa*, many large Malawian and Victorian *Haplochromis* spp., the large *Hemichromis* and *Lamprologus* spp. *Cichlasoma friedrichsthalii*	High-protein pelleted foods; frozen or freeze-dried zooplankton; frozen whole smelt; thin strips of frozen fish fillets; live earthworms; snails; bloodworms; adult *Artemia, Daphnia*, mosquito larvae. Two or three small daily feeds or one large daily meal.
Herbivores *Cyathochromis* spp., Gephyrochromis spp., *Labeotropheus* spp., *Oreochromis* spp., *Petrochromis* spp., *Petrotilapia* spp., *Pseudotropheus* spp., *Sarotherodon* spp., *Simochromis* spp., most *Steatocranus* spp., *Tropheus* spp. C. nicaraquense	High-roughage, low-protein foodstuffs; prepared flaked foods enriched with vegetable matter; algae; fresh vegetable foods, eg balanced baby marrows (zucchini), green peas; weekly feed of frozen bloodworms or glass worms for animal protein; chopped earthworms. Several light feeds daily or two larger daily meals.

offer flake, pellet or foodstick, and frozen or live foods in a planned rota. Some species, such as *C. sieboldii* and *Neetroplus nematopus*, are known to eat algae and other plants in the wild. An aquarium equivalent of this natural diet would be leaf spinach, lettuce and certain vegetable flake foods.

Try to avoid using the powdery remnants of flake food when the tub is almost empty. The fish refuse these small particles and they pollute the system. If possible, buy the 'large flake' brands or 'cichlids flakes' to avoid this problem.

Overfeeding or a breakdown in the bacterial balance of the system can sometimes show itself by the presence of planarians in the gravel and on the glass. These free-living flatworms do not appear to harm fish (although they will destroy eggs and fry) but their proliferation usually indicates an excess of food in the aquarium. Use a copper-based or general anti-parasite treatment and reduce the amount of food for a period of several days to discourage the planarians. Do not make large water changes with raw tap water, however, because this will suppress the natural nitrification processes carried out by bacteria in the undergravel filter. If you do feel a water change will help the situation, aerate stored quantities of water overnight to help dissipate the purification additives.

Below: *A pair of* Cichlasoma bifasciatum *feeds on lettuce. A varied diet is essential for Central American cichlids and some need regular amounts of green food.*

Foods and feeding

There is a long list of nutritionally valuable foods available to the modern aquarist. The strategy in selecting appropriate foods requires knowledge of what and how particular fish eat in the wild. Some species are not finicky and, after a short period of time, will adapt to eating prepared foods. Other species resist all but live foods. Know your fish before buying them, and resist obtaining any species you will not be able to feed properly. In general, fish health depends on your ability to get them eating a *variety* of foods. Foods is not a subject on which to be penny-wise and pound-foolish.

Live foods are often available from your dealer. These include *Daphnia*, adult *Artemia*, glassworms and Tubificid (*Tubifex*) worms. Although the latter are particularly relished, they are collected from sewage and unless purged thoroughly in clean water before feeding, they can wreak

Left: *Some fish, like this pike cichlid, require live 'feeder fish'. The wise aquarist will research his intended fishes' dietary requirements before purchase.*

havoc. In fact, It is unwise to include *Tubifex* worms in the diet of any African cichlids. There is a strong correlation between their frequent appearance on the menu and the incidence of systemic bacterial diseases in cichlids.

Similar problems often accrue from feeding live 'feeder fish' (eg guppies, minnows, goldfish) to large cichlids. Sometimes the feeder fish themselves carry diseases that prove harmful or fatal to the fish that eat them. It is advisable not to use live feeder fish for this reason.

Some South American cichlids, like Uaru or Severum, are partially herbivorous. For cichlids requiring plant material in their diet there are several options. The addition of spirulina or algae-containing prepared foods will provide some of the needed nutrition. However, a better solution is to offer peas, lettuce, spinach or sliced, par-boiled courgettes (zucchini squash). After getting used to these foods, herbivorous cichlids will eat them ravenously.

Discus are renown as 'finicky feeders' although there are other species in the cichlids groups that can be just as fussy towards their diet in captivity. With that factor in mind some of the following points should also be considered relevant for other 'difficult to feed' species.

It is essential that healthy animals receive a broad-based diet of fresh and prepared *nutritionally-rich* foods. Discus, recently imported, or newly introduced to the aquarium will certainly eat live bloodworms etc., which barely keep them alive). They can also be easily encouraged to eat finely chopped earthworms and staple carnivore or cichlids-biased flaked food when kept in the presence of competitive 'dither fish'. Tetras, ideal 'dither fish', will feed by darting upwards in aquaria for surface-floating food flakes. This simple behavioural action can encourage subdued Discus to begin to look for food. Live foods are not always nutritional enough (having been starved of their own food supply after collection and distribution to the aquatic stores) and often will not aid stressed cichlids to overcome disease.

The maxim with Discus is to feed little but often. To ensure the well-being of African cichlids, offer them a diet based on their known nutritional requirements and stick to a fixed schedule of regular feedings. As a useful rule of thumb, offer African cichlids no more food at any one time than they can consume in approximately five minutes.

Regardless of their feeding patterns, many African cichlids need a regular supply of foods that contain beta-carotene and canthaxanthin to retain the full intensity of their coloration. Fresh vegetable foods and frozen zooplankton are good sources of both these substances.

Successful aquarists often make up their own shrimp and green foods to supplement a staple diet of good quality flaked foods. There is an increasing awareness that fish foods needs to be nutritional, palatable and should contain the necessary vitamins and minerals to help maintain fish health. If in any doubt as to the correct manner in which to proceed, it is advisable to talk with experienced aquarists and retailers and discover the best foods available.

Disease is a condition that entails the interaction of a suitable host (ie the fish), its environment (ie the aquarium) and a pahogenic (ie disease-causing) organism. Most fish pathogens are present at all times, but they cannot infect a potential host unless its normal defences have been weakened by environmental stress. Successful disease therapy, therefore, entails more than simply treating the obvious symptoms. If treatment is to prove successful, you must also improve the underlying environmental conditions that allow the pathogen to invade its victim in the first place.

Environmental stress

Abrupt changes in environmental conditions, perhaps caused by equipment failure, such as a heater-thermostat breaking down or a power filter becoming blocked, are well-known stress factors. The resulting flux in water chemistry or temperature are invariably implicated in outbreaks of 'ich' (caused by the protozoan *Ichthyophthirius multifiliis*), a parasitic disease.

Several effective remedies are widely available; keep one on hand in case you need it.

It is always best to treat the aquarium twice or even three times over a 10-day period because cysts may remain in the system that could cause a secondary cycle of infection. Ideally, always dose the aquarium with a suitable white spot treatment when introducing new fish after their quarantine period.

Always try to identify the cause of white spot, especially if it occurs in well-establishes fish. If a heater has failed, fit a new one quickly.

Poor nitrogen cycle management is of even greater significance. Long-term exposure to elevated concentrations of metabolic wastes severely weakens a fish's immune system, making it vulnerable to velvet and systemic bacterial infections, such as haemorrhagic septicaemia and particularly – in the case of African cichlids – 'Malawi bloat'. Behavioural stress can have identical results. Removing sources of stress are, therefore, an integral part of any successful course of treatment. Simply adding

Below: *The typical white spots of ich, or white spot, are clearly visible on this South American cichlid (Geophagus hondae). This condition responds well to proprietary treatments.*

Above: *These pitlike lesions on the head and along the lateral line are characteristic of 'hole-in-the-head' disease. Improve tank conditions and provide medicated foods for one week.*

medication to the aquarium in response to a particular set of symptoms will not solve the problem of disease.

Correcting nitrogen cycle mismanagement may be simply a matter of adhering more rigorously to a schedule of partial water changes. More commonly, it means either reducing the number of fish in a given tank or upgrading its filtration system. Eliminating behavioural stress may also require thinning out a tank's population, but it may be enough to identify and remove an obvious bully, or simply to increase the amount of shelter available.

Parasitic infestations, including ich or flukes, respond favourably to a wide range of proprietary medications. Bacterial infections, however, are much more easily prevented than cured. The prognosis for cichlids so afflicted is guarded, even with antibiotic therapy. When this treatment is not available, recovery is highly unlikely. Under such circumstances, the most humane – and productive – course of action is to dispose of the afflicted fish and concentrate on preventing the spread of the disease to those healthy individuals that remain.

There are several humane methods of destroying a terminally

Below: *A close-up of a single 'ich' parasite,* Ichthyophthirius multifiliis. *Up to 1mm (0.04in) across, with a characteristically curved nucleus.*

Below: *A microphoto of two fill flukes* (Dactylogyrus *sp.*) *clinging to a fish gill. Heavy infestations can cause respiratory distress.*

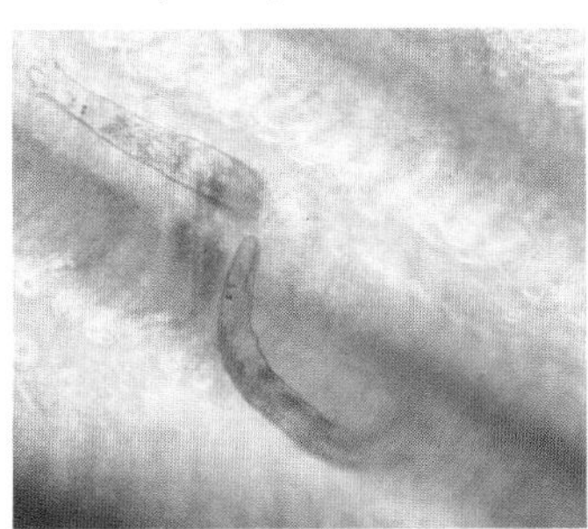

Table of cichlid health

Signs	Possible causes
Fishes itching and scraping body against objects	Excess ammonia/nitrite levels in new aquaria. High pH – new tank (free ammonia in water) Excess nitrates Low pH
Discrete white spots on body and fins	White spot (Ich) *(Ichthyophthirius multifiliis)*
Dusty film on body/fishes gasping at the water surface	Fresh water velvet *(Oodinium pillularis)*
Fishes flicking/gill rate fast	Low pH (3.8-5.5).
Gill rate fast and itching	Water poisoned with insect spray, cleaning agents, etc Gill flukes
Fish gasping (gill rate fast) at water surface	Low oxygen/low pH Freshwater velvet
Cloudy eye, fin rot Blood streaks on fins Sores on body	Poor water quality Overcrowding Inadequate filtration Low pH
Body wounds	Stressed cichlid bullied by dominant fishes
Loss of appetite and hole-like lesions in the head. Erosion of lateral line system	*Hexamita/Spironucleus* (Hole in the head) Poor water quality
Loss of appetite with appearance of clear or white-ish, mucus-like faeces; bloating of the abdominal cavity often follows	'Malawi bloat' commonly the result of bacterial septicaemia

Action
Carry out partial water changes. Check nitrite and ammonia levels. Adjust down to more suitable pH Use ammonia absorber Large partial water change. Monitor nitrate levels
Treat with white spot remedy
Use a proprietary velvet treatment
Extract organic debris from gravel and large partial water change. Monitor pH.
Carry out 90% water change. Use dechlorinator. Switch off filter (6 hrs) and aerate. Use carbon filter. Use 3% salt bath or anti-parasite remedy
Check pH/carry out water changes
Carry out a water change and remove sediment from gravel. Add pH corrector. Add general antibacterial treatment or tonic or antibiotics in separate tank. Check filter and pump size are adequate for stocking levels.
Remove to separate aquarium and use antibacterial treatment
Use antibiotics available on a veterinary prescription. Increase water changes. Check pH, nitrite, nitrate and ammonia levels.
Treat affected fish with minocycline hydrochloride – 250mg/38lt of water. Repeat after 2 days if initial treatment does not lead to resumption of feeding. Feed infected fish with medicated food. Removal of environmental stress is essential for successful treatment of systemic bacterial infections.

ill fish. The simplest is a swift slam against a solid surface, such as the bottom of a bathtub. If an aquarist cannot contemplate such a direct approach, you can dispose of a small fish by leaving it in a cup of soda water at room temperature until there is no trace of gill movement. The dissolved carbon dioxide that causes the soda to effervesce acts as an anaesthetic overdose and simply puts the fish permanently to sleep. A tenfold dose of a commercially available fish tranquillizing agent will have the same effect. This approach is recommended for cichlids 10cm (4in) or larger.

Physical injuries

Treating physical injuries is part and parcel of cichlid husbandry. Most species are aggressive to a certain degree and, in the confines of an aquarium, this tendency is apt to result in split or chewed fins and missing scales or similar skin damage. Fortunately, cichlids have remarkable recuperative powers and, as long as injured fish are given the correct supportive care, they can be expected to recover fully from truly appalling injuries.

Remove injured specimens to a hospital tank containing freshly drawn water of the same temperature and chemical make-up as their original aquarium. Treating superficial injuries, including torn or nipped fins, a few missing scales or small abrasions on the body or head, requires nothing more than the addition of water-soluble vitamin B_{12} to the hospital tank and raising the water temperature by a few degrees to promote healthy tissue regeneration.

Serious injuries, such as fins chewed back to their bases, open wounds or massive scale loss, require more careful attention. Several excellent proprietary remedies are available; be sure to use them strictly in accordance with the manufacturer's instructions. To prevent secondary bacterial infections, dose the hospital tank with a furan-ring

Below: *Poor water quality is often the root cause of fish disease. Regular monitoring can identify water quality problems, enabling corrective action to be taken before it affects the fish. If a fish disease problem is diagnosed, water testing is very important. It helps identify the cause of the problem, making diagnosis easier, treatment quicker, and recurrence much less likely. Good tablet test kits, like the one illustrated, do not require technical knowledge. They are easy to use, and provide simple colour comparison charts to help aquarists determine water quality.*

Above: *Most common fish diseases, including fin, mouth and gill rot, fungus, white spot and other protozoal parasite infestations, ulcers and dropsy, can be successfully treated with products available from aquarist suppliers. Good proprietary treatments are based on years of experience and contain diagnosis charts. Dosing is easy and, importantly, there is no need for aquarists to take the risk of mixing their own chemicals.*

based antibiotic and with a 'liquid bandage', as well as with soluble vitamin B_{12}. This therapeutic regime offers the most hope that tissue regeneration will follow normally and result in minimal scarring or fin ray deformation.

As with accidental poisoning, serious injuries are easier to prevent than to treat. Serious fighting between cichlids is invariably the result of insufficient living space, a shortage of shelter, or a breakdown in the normal reproductive sequence. If you familiarise yourself with the space requirements of a given species before adding it to the tank and follow the suggestions outlined on breeding cichlids (see page 51 to 61), you should not, as a rule, find aggression an unmanageable or insoluble problem.

Signs of disease and remedies

The best way of avoiding diseases in the first place is to buy healthy fish from a good aquarium dealer who *clearly* maintains high water quality in his aquaria. Ask penetrating questions about the requirements of the fish *before* purchase. A good dealer will always provide answers.

The hobbyist is often dealing with wild-caught specimens. Often they are emaciated and pick at, but reject, all offered foods, live or otherwise, and produce white stringy faeces. This syndrome may result from any of the following organisms: the protozoans *Hexamita* or *Spironucleus*, or threadworms (*Capillaria*), or others. Metronidazole is the drug usually recommended for the treatment of flagellate protozoans in cichlids and Dylox (Masoten) for threadworms. Naladixic Acid, an antibiotic has also proven exceedingly useful. The use of specialized drugs for specific diseases requires care and attention to detail. It is advisable to consult up to date publications or seek professional advice. Finally, wild fish are often parasitized with gut worms. There are a number of medicated prepared foods on the market that pack several antihelminthics in palatable form which may be useful in 'worming' fish.

So-called 'hole-in-the-head' disease (caused by the single-celled parasites) *Hexamita* and *Spironucleus*) can affect cichlids in overcrowded aquarium systems or in conditions of poor water quality. A proprietary treatment is available to treat *Hexamita* and specific anti-protozoan drugs, such as metronidazole and dimetridazole (for which a veterinary prescription may be necessary) may also be effective. Some fishkeepers advocate using antibiotics to treat 'hole-in-the-head' disease. All these treatments should be administered to the affected fish in a quarantine aquarium. In practice, the success rate for cichlids is rather poor once the disease manifests itself, but few other options exist.

The majority of bacterial and parasitic infections are wrongly identified by aquarists, who consequently apply an inappropriate treatment to the aquarium. The true fungus (*Saprolegnia*), for example, rarely occurs. The general white tissue damage sometimes attributed to fungus is usually caused by a bacterial infection. Always check the signs and make the correct diagnosis before taking any action. When in doubt, ask experts.

Many books refer to gill flukes (*Dactylogyrus* and other species). Once a fish begins to itch and scrape its body on rocks, gill flukes are invariably suggested to be the cause. In fact, gill flukes are likely to be at the bottom of the list of possible causes. Here again, it is vital to rule out the most likely causes before blasting the aquarium with the wrong treatment. Healthy fish rarely succumb to disease if the water quality, stocking levels and feeding programme are correct.

Newly imported discus are prone to infections. It seems these cichlids are especially stressed by poor water conditions and overcrowding during export/import and distribution through the fish trade. Young discus transport better than large adult and wild discus.

It is important not to mix discus from different sources unless there is complete confidence in their health status. It has been recorded that farm-raised discus from the Far East may have different immunities and strengths than American or German forms. Introduction of either into a community can quickly lead to cross infections. Some specimens have been known to carry and are vulnerable to TB infections.

As is common with all cichlids, fully established captive discus rarely develop infections unless there has been a regime of an inadequate diet and poor water quality and they are allowed to prevail. Low stocking levels and/or inadequate aquaria promote exaggerations in pecking orders which can lead to over-aggression within a group.

A darkening of the body colour, white faeces and 'hole in the head' suggest that at some time during import/export/wholesale/retail and time in captivity, excessive pressure has been brought to bear relating to the three important factors of water quality, diet and husbandry in terms of stocking levels and peer size within a discus community.

Discus and Eartheaters

Finally some species, in particular Eartheaters, seem prone to the 'Neotropical Bloat' syndrome, in which the affected fish will suddenly develop pronounced abdominal swelling, begin ventilating heavily, stop eating and die after hanging listlessly at the top of the tank for several days. In contrast to 'Head Hole' or neuromast pitting, lax water maintenance seems not to be the cause nor is the condition contagious. Nothing seems to cure it.

When dealing with a sickly discus the very first step should be to isolate the stricken fish from the healthy group. The sick discus should then receive treatment in a separate quarantine aquarium specially set up and containing healthy and active dither fish. If allowed to remain with a group of discus a diseased fish of the same species would be bullied and further stress inflicted.

To make matters worse, if quarantine measures are not taken immediately then the diseased fish could possibly pass on the infection to other fish in the aquarium.

Below: *'Head Hole' or neuromast pitting in cichlids is often caused by lax water maintenance. Take care that a high level of water purity is maintained at all times.*

AFRICAN CICHLIDS Dr Paul V Loiselle

Monogamous African cichlids
Most representatives of this group are substrate-spawners, but it also includes both primitive and advanced mouthbrooding species. What sets these cichlids apart from the two polygamous groups is the pair bond. This behavioural mechanism allows two aggressive – and often individually territorial – fish to share the same space without incident for the duration of the breeding period. Success in breeding these cichlids depends entirely upon the formation and maintenance of a pair bond.

Few aquarists simply select a male and female at random, drop them together into their own tank and expect pairing to follow automatically. To begin with, both fish need to be sexually mature, compatible and willing to breed. The presence of target fish to serve as an external focus for the aggressive behaviour of the prospective pair also plays an important role in sexual conditioning. Rearing to maturity a group of six to eight juvenile fish of the same species is the most reliable way to obtain a compatible pair of monogamous cichlids. This approach works well precisely because it offers prospective pairs both a choice of partners and plenty of suitable targets for their aggression. The same principle explains the ease with which a single male and female often pair up in a community setting. However, both approaches entail an element of risk to the other fish in the tank. Indeed, unless the aquarium is large enough to allow tankmates to escape the attentions of the newly established pair, it usually proves necessary to remove them for their own safety! These surplus fish can be encouraged to pair off in turn, or kept in a community situation, or sold to other aquarists.

Target fish
The presence of 'enemies' also plays a role in maintaining the pair bond, although its importance varies greatly between all cichlid species. The lamprologine cichlids of Lake Tanganyika have remarkably strong pair bonds that persist even in the absence of any target fish. The true tilapias, on the other hand, seem incapable of maintaining a pair bond in isolation. Most monogamous species fall somewhere between these two extremes.

The simplest, although not always the most practical, means of providing sufficient external stimuli to sustain the pair bond is to set up a pair in a tank large enough to afford other fish a refuge beyond the limits of the breeding territory. This may entail doubling the minimum recommended bottom area of the breeding tank, so the feasibility of this option is in inverse proportion to the size of the cichlids in question! Fortunately, the mere sight of potential territorial rivals or spawn predators is enough to reinforce the pair bond. Placing the breeding tank next to a well-stocked aquarium, or isolating a target fish behind a clear partition in the breeding tank serves this purpose satisfactorily.

The breeding tank
Some cichlids are so aggressive that it is impractical to allow the two sexes unrestricted access to one another. In sexually dimorphic species, the easiest way to assure the female's safety is to separate the sexes with a barrier containing openings through which she can pass freely, but which restrict the movements of the larger male. Place the spawning site in the male's territory, which should be the larger area of the two. This 'separate compartment' approach gives the aquarist an opportunity to observe typical reproductive behaviour with a minimum risk to the female, since it allows her to initiate contact with the male, as well as to break it off should he prove too rough a suitor. It also

affords her a secure refuge should the pair bond break down after spawning has occurred.

The risk of death or injury to the female can be eliminated completely, albeit at the cost of the normal expression of the fish's reproductive behaviour, by separating the sexes with a divider that permits free circulation of water between the two compartments of the breeding tank. This 'incomplete divider' approach exploits the fact that a male and female cichlid will perform their respective roles in the spawning sequence without physical contact, as long as they are in sight of one another. Assuming there is a spawning site immediately adjacent to the barrier and good water circulation in the breeding tank, sufficient sperm will diffuse through the grid to fertilize a significant percentage of the female's eggs. Once the fry are free swimming, they can move freely between the two compartments, thus allowing each adult the opportunity to fulfil a parental role.

Although they differ over what constitutes a suitable spawning site, all these cichlids initially prefer a free choice of sites. Indeed, the selection process, no less than actual site preparation, appears to play an important role in reinforcing the pair bond in nature. Established pairs often develop a marked preference for a particular spawning site and when this happens, the breeder can adjust the tank furnishings accordingly.

A factor complicating the husbandry of some of the monogamous cichlids native to forest streams is that the sex ratio of their progeny is determined by the pH and hardness of their surroundings. This would not matter greatly if these fish, all representatives of the genera *Pelvicachromis* and *Nanochromis*, spawned only when pH and hardness values fell within an optimum range. Unfortunately, these cichlids willingly spawn over a much broader range of pH and hardness values, and this results in unisex broods. This peculiarity of their natural history is probably the chief reason why so few *Pelvicachromis* species have become commercially established, notwithstanding the relative ease with which wild-caught fish can be induced to breed in captivity.

Premature respawning

Once the pair has spawned, the fishkeeper need usually do no more than provide food for the free-swimming fry. In the wild,

Below: *A 'separate compartment' tank is ideal for breeding isolated pairs of monogamous cichlids, such as* Hemichromis guttatus. *The openings in the divider allow the smaller female access to the male and a means of escape if he becomes too aggressive. Smooth stones offer a choice of spawning sites; floating plants provide cover.*

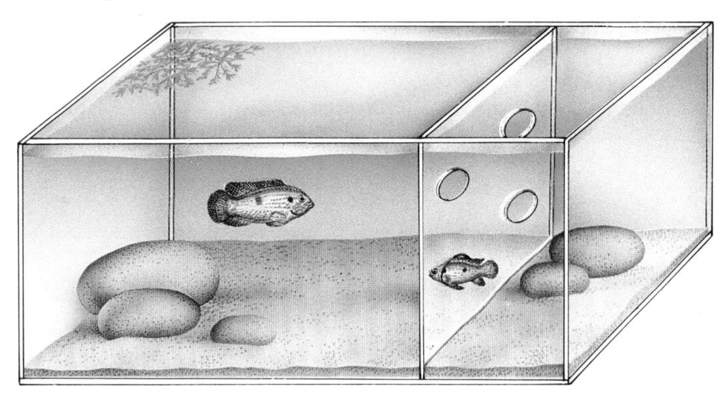

effective custodial care places such demands on the time of the parent cichlids that most are unable to forage normally. They must thus subsist to a significant degree on body fat until their fry grow large enough to forage beyond the limits of the pair's original territory. This constraint on normal feeding precludes the possibility that the female will ripen another batch of eggs while tending fry. In captivity, thanks to the fishkeeper's generosity at feeding time, adult cichlids have no difficulty ripening a second batch of eggs while tending young from an earlier spawning.

Lamprologus brichardi and a number of *Julidochromis* species, are not restricted biologically from spawning again while still caring for their first brood. These cichlids have also evolved behavioural mechanisms that permit the adults to cope with several broods of fry within their territory. For most monogamous cichlids, however, such a situation can lead to serious problems, as one or both adults tend to lose interest in the older fry with the onset of a new spawning cycle. The pair will often attempt to drive their first batch of fry out of the breeding territory as a prelude to respawning, regardless of the brood's ability to survive on its own. If the young cannot move beyond the limits of the breeding territory, they are likely to be killed by their former guardians. In some instances, one parent – usually the female – will attempt to drive the older fry away, while the other parent attempts to defend them. This can lead to serious intersexual fighting and such premature respawning efforts doubtless account for many otherwise inexplicable breeding failures.

The simplest way to prevent such an eventuality is to reduce the temperature in the breeding tank to 22-23°C (72-74°F) as soon as the fry become mobile. This will slow their growth but, more importantly, it will also retard the maturation of a new clutch of eggs. Under such conditions, it is far more likely that the parents will tend their fry until they become fully independent. If you opt for higher tank temperatures and faster fry growth, be prepared to separate parents and offspring at the first signs of respawning, such as overt courtship behaviour or preparation of a spawning site.

Harem polygynists

The chief characteristic of fish that adopt this mating system is that a single male monopolizes a group of two or more females, each of which defends a separate breeding territory against others of her sex. All the African representatives of this group are cave-spawners. Care of the eggs and yolk-sac fry is an exclusively maternal task, but male participation in the defence of the mobile young is not unusual and may be solicited by the female.

While there is a bond of sorts between a male and the members of his harem, it does not compare in intensity with the pair bond of monogamous cichlids. Females seem to accept any male as a spawning partner, while males will attempt to spawn with as many females as they can. Indeed, the more females present, the less danger his attentions pose to any one of them.

Breeding tank size is the most significant factor to influence reproductive success of haremically polygynous African cichlids in captivity. Obviously, it must be large enough to afford each female a territory of her own but, equally important, it must be large enough to afford a secure refuge for the male, whose proximity to the spawn is not appreciated during the initial phase of the reproductive cycle.

Openly polygamous African cichlids

Representatives of this large assemblage of species lack the behavioural mechanisms that permit long-term cohabitation of the sexes. Contact between male and female is limited to the actual spawning act, and it is not unusual for an individual of either sex to have multiple partners during the course of a single spawning effort. All openly polygamous African cichlids are maternal mouthbrooders, the male playing no role whatsoever in the care of the young.

Among the majority of openly polygamous species, the male defends a discrete breeding territory. The exceptions to this rule are the several *Tropheus* and *Cyprichromis* species of Lake Tanganyika and a number of Malawian *Haplochromis* of the open-water dwelling utaka group. Males of these non-territorial species actively attempt to separate a ripe female from other males, rather than lure her into a fixed territory. The persistence of territorial behaviour, no less than the size and spacing of breeding territories, varies considerably between representatives of this group in nature. These differences are less important in captivity, however, where space limitations restrict the expression of much species-typical behaviour. A single management approach, therefore, works equally well for all these cichlids under normal aquarium conditions.

Openly polygamous cichlids are best maintained either in groups of a single male with several females, or as single pairs in a community setting. Either approach results in an environment where the male is too preoccupied to harass a female to the point of killing her – the usual outcome if these cichlids are housed as isolated pairs. Serious breeders favour the first approach, as it maximizes fry production.

Aggression in these cichlids is intimately linked with reproduction. Thus, while it is unwise to try housing several males of the same species in an aquarium containing

Below: *These open-spawning* Hemichromis guttatus *share the defence of their breeding territory and protect fry against predators.*

females, more than one male of the same species will live amiably in a bachelor tank. Fish of the same species and sex are the most obvious targets of a territorial male's aggression, but it is not unusual for males of another species with a similar colour pattern to come in for unwanted attention as well. Consequently, it is not a good idea to house species with very similar male breeding dress together. A sexually active male grows progressively more belligerent towards tankmates, regardless of their identity, as his potential consort ripens her eggs. However, once spawning is over, the intensity of his territorial behaviour declines dramatically. In a community set-up, males will typically succeed one another as the tank's dominant resident, as their respective females attain reproductive readiness.

Once spawning has occurred, it is a good idea to remove the female to a nursery tank to complete the incubation sequence. As long as the main breeding tank contains sufficient cover to allow her to avoid immediate post-spawning harassment by the male, there need be no rush to isolate the female. With rare exceptions, maternal mouthbrooders are exemplary mothers, even under crowded conditions. However, isolation is essential if you wish to save any of the fry. While females will carry to full term in a community setting, they can seldom provide an effective defence of their mobile young in such an environment.

Provide a nursery tank in scale with the size of its intended occupant. A 20 litre (4.4 Imp. gallon/5.3 US gallon) tank will suffice for females up to 7.5cm (3in) long; a 38 litre (8.4 Imp. gallon/10 US gallon) tank for fish measuring 10-15cm (4-6in); a 57 litre (12.5 Imp. gallon/15 US gallon) tank for fish 15cm (6in) and larger. Fill it with freshly drawn water of same chemical make-up and temperature as that of the spawning tank. Add a thin layer of gravel, as the light reflected from bare glass is very disturbing to these cichlids. The tank should contain shelter of some sort – a flowerpot or two facing away from the front glass will suffice – and a layer of floating plants. A mature sponge filter, a heater-thermostat and cover complete the furnishings.

Contrary to popular belief, many egg-carrying females will feed at reduced levels during the incubation period, if the opportunity presents itself. Supplementary feedings are not essential to their well-being and complicate nitrogen cycle management in the nursery tank. However, cessation of normal feeding does not mean that the normal metabolic processes also cease. Brooding females and their developing young generate significant quantities of nitrogenous wastes, therefore continue to make small but frequent water changes throughout the incubation period.

The fertilized eggs remain in the throat sac of the female for 10-28 days, depending upon the species in question and on the water temperature. At the end of this period, the female releases the fully developed young. In a few species, such as the *Labidochromis* of Lake Malawi, the fry then fend for themselves, but in most cases, the female continues to defend the free-swimming fry as they forage and will allow them to re-enter her mouth when seriously threatened.

Such prolongation of maternal care doubtless contributes significantly to fry survival in nature. However, in captivity it can actually threaten their well-being. Females often refuse to release their fry into what they perceive as unsafe surroundings, or they may be so nervous that even if they do release the fry they take them up again at the slightest disturbance. This overprotectiveness seriously interferes with the normal feeding behaviour of the fry. Thus most aquarists terminate the relationship between mother and young following their initial release.

Separating mother and fry is a simple matter in the case of females measuring 7.5cm (3in) or larger. Simply enfold the female in a net, hold her head-down over a container of water taken from the nursery tank, gently pull open her lower jaw and immerse her head in the water. The fry will immediately swim out of her open mouth. Females less than 7.5cm (3in) long are more difficult to manipulate safely. The following procedure sounds Draconian, but is less stressful to both mother and young than more direct forms of manipulation. Insert the female head-down in the tube of a kitchen baster. Replace the bulb and insert the lower third of the baster in a container of water from the nursery tank for a few moments. Remove the baster from the water for a minute or two, then replace it and give the bulb a few good squeezes. If no young are expelled from the baster's opening, remove it from the water for a few minutes and repeat the procedure. Respiratory distress will eventually force the female to make repeated gulping motions. The reverse flow of water produced by squeezing the bulb works with the female's jaw action to squirt the young harmlessly into the container.

Many breeders advocate a reconditioning period for the female before she is returned to the breeding tank. However, unless she is severely emaciated at the end of the incubation period, this is not essential. The fishkeeper can shake up the pecking order in the main tank before reintroducing the female by rearranging the tank furnishings and carry out a large-scale water change. The female is thus spared serious harassment from her tankmates.

Artificial incubation of cichlid embryos

As a rule, cichlids are excellent parents. Nevertheless, it may transpire that a pair consistently devours its spawn, or that a female mouthbrooder fails to carry to full term. In most cases, the culprits are young, inexperienced fish. Exercise a bit of patience and the problem will often correct itself. However, if the fish are extremely rare or valuable, opt for a less passive response to such behaviour. It is possible to incubate cichlids zygotes (fertilized eggs) artificially, although there is a price to be paid in developmental casualties and congenitally deformed fry.

Artificially incubating the zygotes of the substrate spawning cichlids

Separating a small female mouthbrooder from her fry

1 Insert female head down in the tube of a kitchen baster.

2 Insert lower third of baster into container of water from nursery tank. Remove baster from water.

3 Replace baster and squeeze bulb. Young are safely expelled into the water.

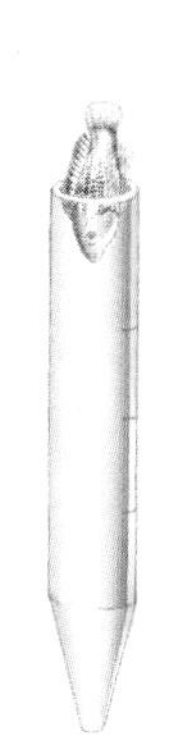

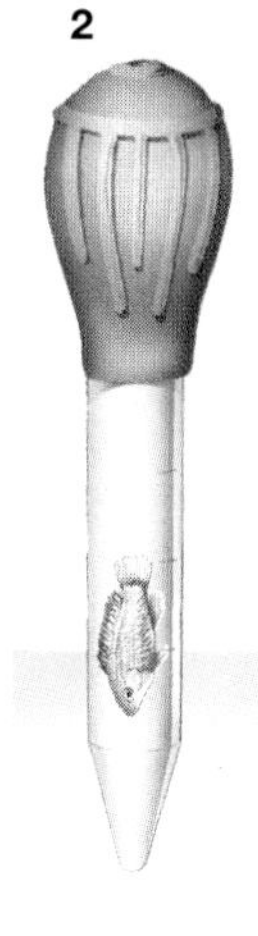

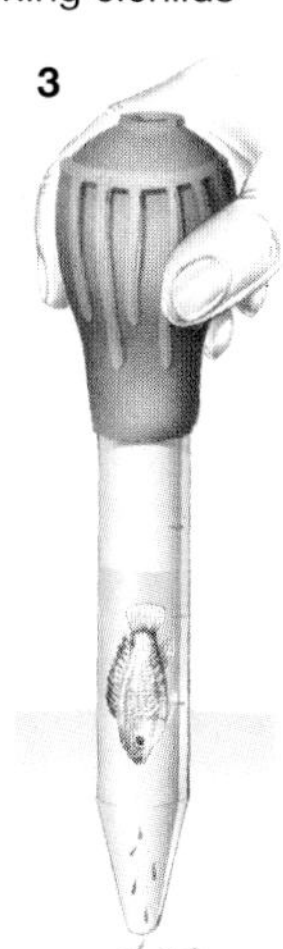

is a relatively straightforward matter. Transfer the leaf or rock on which the eggs were laid to a small aquarium filled with freshly drawn water of the same temperature and chemical make-up as that in the breeding tank. In order to duplicate the fanning behaviour of the parent female, place an airstone a short distance from the clutch and regulate the air flow to produce a moderate stream of bubbles. The bubbles must not come into contact with the spawn, nor should the current they produce be so strong that the developing young are battered by its flow. Failure to observe these conditions will result in a substantial percentage of congenitally deformed fry. The zygotes of cave-spawning species are light sensitive, so it is a good idea to cover the sides of their hatching tank with brown paper until the young become free swimming. A thermostatically controlled heater and secure lid are enough to complete the hatching tank's furnishings.

The developing young are vulnerable to bacterial attack. To minimize losses from this source, treat the water in the hatching tank with an antibacterial agent. Methylene blue is traditionally used for this purpose, but it oxidizes so rapidly that you must repeat the dosage several times while the eggs are developing if it is to retain its potency. Proprietary solutions of acriflavine hydrochloride, used according to the manufacturer's instructions, are a superior alternative to methylene blue. Satisfactory results can also be obtained with medications based on stabilized chlorine oxides.

Once all the fry have hatched, it is a good idea to change half the water in the tank. Repeat this procedure as soon as they become free swimming. Once the

Below: *Young pairs of kribensis may eat their first spawns, but in time become excellent parents. Once the female brings the young out, both adults tend the fry.*

young have reached this point in their development, they should be treated in the same manner as fry left under parental supervision.

Artificial incubation
Artificial incubation of mouthbrooder zygotes is a more difficult proposition. A successful 'artificial mouth' must incorporate some system of gentle water circulation that mimics the tumbling the zygotes receive within the maternal throat sac. However, the current must not be so strong that the young are dashed against the sides of the container; such treatment will either kill them outright or cause serious congenital deformities. By far the easiest way to produce such conditions is to gently aerate the contents of a small container suspended in the heated water of a hatching tank to assure thermal stability within it.

It is essential to use some sort of antibacterial agent in the 'artificial mouth'. The preparations recommended in the artificial incubation of substrate-spawning cichlid zygotes are just as effective in the case of mouthbrooder fry. Bear in mind that in some instances, the dosages recommended by the manufacturer may prove harmful to mouthbrooder zygotes, so it is advisable to cut the recommended dose by half at the first attempt. If a higher concentration seems in order, increase the dose on subsequent occasions.

On alternate days, replace half the volume of water in the 'artificial mouth' with fresh water of the same temperature and chemical make-up. Remember to bring the concentration of antibacterial agent back up to strength following each water change. Check the progress of the zygotes frequently and remove any dead individuals from the 'mouth' as soon as you notice them. If they are left where they are and are allowed to decompose, further losses among the fry are unfortunately inevitable.

The percentage of mortalities depends primarily upon the state of development of the zygotes at the onset of artificial incubation. The more time they have spent within the female's mouth the greater their chance of survival. Older zygotes are also less susceptible to congenital injury than unhatched eggs or youngsters at a very early stage of development. Once the young are fully mobile, they can be gently decanted into the hatching tank and offered food. For the aquarist, their rearing from this point on is a straightforward proposition.

Artificial incubation of mouthbrooder zygotes is a time-consuming process and the yield of normal fry is invariably inferior to that resulting from normal maternal incubation. It is thus surprising that many breeders routinely 'strip' a female as early as three days after spawning to put her zygotes into an artificial mouth of their own devising. On grounds of convenience alone, human intervention in cichlids' brood care is best regarded as a last resort, rather than a routine matter.

Rearing African cichlids fry
Cichlids fry are easily reared. Those of *Anomalochromis thomasi* and some of the dwarf lamprologines require microworms for their first few meals. As soon as they are free swimming, the fry of other African cichlids can manage brineshrimp nauplii and finely powdered prepared foods. As improbably as it may seem, even adult fish of robust species will snap up newly hatched brineshrimp with gusto. Always take this adult competition into consideration when calculating the quantity of brineshrimp nauplii to offer fry under parental supervision.

Because of their small size, cichlid fry require a minimum of three feedings daily to prosper. Tanks that support a dense growth of green algae on their back and sides, or those that contain a fully matured sponge filter, offer the young cichlids a rich and natural supplementary food source. It is not surprising then that fry reared in such a setting grow more rapidly and evenly than those that rely exclusively on their keeper's offerings of food.

Rigorous attention of aquarium hygiene is essential to the successful rearing of cichlids fry. They are extremely intolerant of dissolved metabolic wastes, yet the by products of their hearty appetites do nothing to simplify the task of sound tank management. Nitrogen cycle mismanagement can result in the loss of entire broods that succumb either to nitrite poisoning or to opportunistic diseases that attack the young when their immune systems have been compromised by chronic low-level exposure to dissolved metabolites. The tank sizes suggested earlier for a single breeding pair or harem will usually afford adequate growing room for a brood of fry for 4-6 weeks after hatching. To sustain a satisfactory growth rate beyond this point of development, move them to larger quarters, or if practical for you divide them between several rearing tanks.

Above: *A striking collection of Tanganyikan cichlids that includes yellow* Lamprologus leleupi, L. tretocephalus, Tropheus duboisi *and* t. moorii. Lamprologus *fry grow relatively slowly;* Tropheus *fry grow more rapidly.*

Fry and water changes

By increasing the frequency of water changes, you can delay relocating the fry for a short time. However, not all species respond equally well to such a regime. In any event, as fry tend to be more sensitive to large-scale water changes than their parents, it is best to change a smaller volume of the water in their tank – between 10 and 25 per cent every 2-3 days – than to make a 40-60 per cent change every week. Young fry, in particular, are less tolerant of dissolved chlorine than are larger specimens of the same species, so it is a good idea to use a commercial dechlorinating agent when making water changes in their tanks.

With the exception of the banded jewel fish and the larger, piscivorous *Lamprologus* species, African cichlid fry are less given to sibling cannibalism than most of their New World cousins. The smaller brood sizes of most species also make it easier to provide the fry with adequate growing room. One characteristic of the species bred to date is that male fish grow more quickly. Bear this in mind when culling large spawns or choosing future breeding stock. If only the largest fry are selected for rearing to maturity, the likely outcome of the selection process will be a tank full of males!

CENTRAL AMERICAN CICHLIDS David Sands

Aquarists wishing to breed Central American cichlids must sometimes have an inner desire to attempt the difficult. Nevertheless, many species are regularly spawned and raised by dedicated enthusiasts in the USA and Europe. Other spawnings are often accidental in a community of cichlids or other tropical fish. The high degree of parental care shown in all species can often guarantee some fry will survive even in the most intensely populated aquarium.

Some species are so large, however, that only massive aquaria can accommodate the breeding territories and the defensive roles they play. In small aquaria, a breeding pair can disrupt all other occupants and even kill a couple in the process. Such is the strength of instinct to protect the offspring that sometimes a spawning pair can be the only fish to survive the event!

Above: *A pair of* Cichlasoma cyanoguttatum *tending fry. Once established, breeding and raising fry is a natural progression.*

Selecting a breeding pair

Sexing most cichlids is a difficult affair, especially when they are in the juvenile or semi-adult stages. Since few breeding adult pairs can be purchased without a bank loan, juveniles are the most likely breeding stock for most fishkeepers. It is fairly certain that a pair will develop among a group of juveniles, although if the selection is made from one batch, inbreeding could result by the cross of brother to sister. The best way of avoiding this is to buy the same species from different sources.

It is almost impossible to select

Below: *Jaw locking between an Oscar (left) and a Jaguar Cichlid. This is usually a test of strength between rivals.*

two fish from a batch, place them in an aquarium and expect them to accept each other and spawn. It is quite possible, however, to study a group of cichlids and observe a dominant individual and presume this is to be a male. Then, a less sturdy fish with shorter fins could be selected as a possible female. If they are juveniles, it is best to isolate them in separate aquaria where each can mature. Many fish are lost by being introduced in 'pairs'; the dominant fish almost automatically begins to attack the subdominant one. This is the problem of an enclosed system.

In their natural habitat, semi-juveniles exist in quite large shoals in relatively spacious expanses of water. It is almost impossible, therefore, for an individual to dominate another and cause its eventual demise. In captivity, the dimensions of an aquarium may simply mark out an arena in which a weak fish is killed by an aggressor. Such arenas do not occur in the natural environment; a dominant fish will simply chase off an intruder. In an aquarium, a beaten fish often has nowhere to go to escape attack.

Below: *A Convict Cichlid,* Cichlasoma nigrofasciatum, *having spawned inside this conveniently placed pot, cares for the eggs until they hatch.*

Since good fishkeepers do not wish to witness unnecessary carnage in their aquaria, it is important not to place fish in this situation. Always buy a group of four to six fish and make sure you can accommodate them at their eventual size. Most aquarium dealers will exchange well cared for cichlids that have outgrown their original accommodation.

Spawning and rearing

In an aquarium in isolation, breeding cichlids and raising fry is simply a natural progression.

When keeping cichlids for a reasonable period of time, it becomes obvious when a pair are setting up a spawning sequence. Pairs not yet sexually mature will enter into practice rituals, which usually involve defence of an area of rocks or general debris. The pair will mouth and scrape clean areas of rock or bogwood, preparing the surfaces that the eggs will adhere to. They will circle each other, the male extending his fins in much the same way as a displaying male bird. Sometimes, jaw locking will occur. It is not exactly clear why such head on tests of strength.

If the pair are mature enough to spawn then the female will make runs along the pre-cleaned surfaces and eventually begin to deposit groups of eggs. The male follows the female's run and

fertilizes the eggs. The pair will perform this exercise repeatedly, sometimes over a long period of time, until all the eggs have been deposited. Then both fish will fan the eggs, using their ventral and pectoral fins to sweep water over them.

At this point, immature or unsuited pairs sometimes consume the eggs or take little interest in them, thereby laying them open to predation by other fish. On certain occasions, the pair will argue and a fight between them will end up with a badly damaged female. By contrast, a balanced well-suited pair will develop a strong bond and defend the eggs as their first priority. In this instance, any cichlids or other fish foolish enough to encroach upon their territory will be driven off with a fervour rarely matched in the natural world.

After several days of close attention to the eggs, fry will begin to emerge in small batches until all the fertile eggs hatch. During this period, some aquarists advocate adding a small quantity of a suitable antifungal agent to the water. Use these products – which contain such substances as acriflavine, methylene blue or malachite green – according to the maker's instructions. When treating eggs, always underdose the aquarium to be on the safe side. Such treatments help to combat fungus that invariably develops on infertile eggs and which may affect fertile ones. To some extent, the parents take care of this problem by picking off any infertile eggs from the batch. Once all the fry are free of the egg cases, the parents usually place them in an area in which they can best be defended and cared for.

If the aquarium has a gravel substrate, for the next few days the fry are placed in hollows or pits dug by the parents. Within 48 hours of being free swimming, after the yolk sac has been consumed, you can feed the fry on freshly hatched brine shrimp, microworms or liquid and powdered fry foods. (The brooding parents often offer sustenance to their fry by blowing out a cloud of chewed food.) During the next 14 days offer the fry food on a 'little but often' basis.

When fry have begun to forage for food away from the protection of their parents, remove them to a raising aquarium. Opinions differ on this strategy; some aquarists recommend leaving some of the fry with the parents so that the normal cycle of instinctive care continues in the natural way. Certainly, removing all the fry can sometimes cause aggressive disputes to break out between the

Below: *A female* Cichlasoma hartwegi *in brood colour display, caring for fry. Parents often spit out food for their brood.*

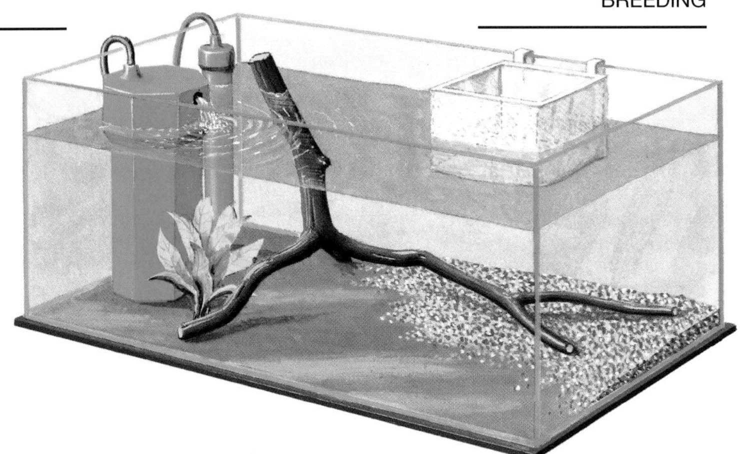

Above: *A fry-rearing set-up suitable for Central American cichlids. Ensure that the system is well established (in the filter sense) before adding any fry. Newly hatched fry can be fed more easily in the net and, once grown, released into the tank.*

parents, and the female may be seriously injured as a result. Conversely, in some instances, fry can remain with the parents too long, leading to problems developing between the parents when another spawn cycle begins. It is not unusual for the parents to lose interest in their brood after a period of about three or four weeks, which suggests that removing the fry from the aquarium at an earlier stage is advisable.

Ensure that a raising aquarium is well established before transferring fry to it from the main tank. Use a 50/50 blend of new and main aquarium water in the nursery tank and install an internal power filter, sponge or box filter. These simple filters are ideal because they do not draw food particles away from the fry as would happen among the gravel particles of an undergravel filter. In fact, sponge filters actually provide a useful feeding surface for the fry, since they can pick off fragments of food that collect on the exposed filter body. Use these filters in a tank with a bare glass base so that any excess food can be removed easily with a siphonic device.

The larger the raising aquarium, the better the growth rate will be. Without doubt, the most important factors that affect growth rate are related to the size of the aquarium, the amount of food and the frequency of feeding. If a large raising aquarium is not practical then frequent partial water changes will help to alleviate the problems of overcrowding fry.

It is not unusual for whole batches of fry to be lost, even when everything seems to be going well.

The breeding sequence described above applies to a spawning pair of cichlids in a separate aquarium. Should a pair spawn in a community system of cichlids and other species, then the pattern is much the same. The egglaying rituals and protection of the spawn will go on despite a crowded aquarium. Developing fry will group close to their parents and will be suitable protected, but predation will undoubtedly occur once the fry begin to stray. This is a normal pattern that simply reflects the universal laws of natural selection.

In a community aquarium, take care to choose suitable species for inclusion with the cichlids. Many, if not all, catfish, for example, are nocturnal and active night feeders. Cichlids will slow down their body system at night and therefore will not be capable of protecting fry just when predatory catfish are most active.

SOUTH AMERICAN CICHLIDS Dr Wayne S Leibel

The majority of South American cichlids form stable monogamous pair bonds through interesting and elaborate behavioural rituals and exercise diligent, even gentle, care of their eggs and youngsters. These courtship and spawning behaviours are remarkable to witness and a wonder to behold and are one of the main reasons for keeping neotropical cichlids in the home aquarium.

Sexing cichlids and establishing compatible pairs

The first step in spawning any fish successfully is obtaining males and females. In the case of South American cichlids, this also means obtaining compatible males and females, since pair bonding and co-operation is so central to reproduction in these fish, at least for the monogamous species which are in the majority. Harem sp polygamous species, where male-female interaction is brief and males are likely to spawn with several females in turn, pose less of a problem: a mating group of predominantly females must be provided with space and shelter.

Many South American cichlids are sexually dimorphic – males and females are easily discriminated on the basis of size, overall shape, finnage or coloration. But some are sexually isomorphic, that is, there are no apparent differences between the sexes. Nevertheless, many 'isomorphic' species may still be sexed on the basis of subtle differences. In general, males of most species are slightly more elongated as measured by the distance along the ventrum from the point of insertion of the paired ventral fins to the anal fin. In general, ripe females are slightly heavier through the abdomen than their consorts. However, these are differences that often only experienced aquarists can detect. The urogenital vents of males and females differ (the female's housing a blunt ovipositor, the male's a pointed tube) and in theory, they can be sexed by netting them and examining the aperture, but in practice, 'venting' is difficult and often unreliable. The proof occurs only when the tubes actually descend during spawning

Below: *Many South American cichlids, like these marbled pike cichlids,* Crenicichla marmorata, *are sexually dimorphic. The female, on top, has a white-edged dorsal fin and develops a cherry-red distended belly when ripe.*

Above: *Some South American cichlids, like these* 'Cichlasoma' atromaculatum, *are sexually isomorpic and not easily sexed. The female of the pair, in the forefront, is slightly more rotund than her consort.*

and the 'egglayer' is identified firsthand. For sexually isomorphic species, aquarists should purchase a group of 4-8 individuals, raise them, and allow them to pair off naturally. As females often grow more slowly and attain a slightly smaller adult size, a good mix of sizes, both large and small, is recommended in that founding group. This is good general advice for obtaining breeding stock of all South and Central American cichlids, regardless of their individual sexability.

Often, it is not possible to pick out several juveniles, rather only one or a few adult specimens are available, or the price prohibits purchasing more than two. For sexually dimorphic species, in general, adult males tend to be larger and to have more filamentous finnage, in addition to whatever coloration differences (sexual dichromatism) exist. These differences are described in the species section for many species, and in some cases are so pronounced that the two sexes in question could be (and often have been) described as separate species! Unfortunately, newly-imported wild specimens are often emaciated, with tattered fins and no colour.

Sometimes, if you are patient and watch the fish in your dealer's tank, their interactive behaviour will suggest their sex. If two cichlids engage in lateral displays or jaw-locking, or one butts the other repeatedly in the genital region, you may be watching courtship behaviour. When all else fails, take a gamble and hope the odds are with you: buy as many as your budget allows and hope the fish will sort it out in your tank.

Assuming both sexes are available, the next task is to establish a compatible pair. The chances of this happening are increased if the two fish are added to a new (ie strange) tank together, so that neither has proprietary rights.

Target fish

As with African cichlids, the use of target fish to help cement pair bonds is a useful strategy in breeding monogamous neotropical cichlids. Target fish are usually other cichlids, either a third

conspecific, or multiple individuals of another species of similar temperament. Larger non-cichlids can often be used as well. The strategy is to provide a reason for the intended pair to work together and vent their aggression towards the other fish and not towards each other. Obviously, appropriate sized tanks with the right type of shelter should be chosen if this strategy is to work properly: death of the target fish should be avoided and is not the desired outcome of this strategy.

It is often useful to separate the intended consorts for a while before they are put together. This allows them to ripen with good feeding and to 'meet' each other visually and chemically before the 'date' is arranged. They may be placed in separate tanks next to each other so they can see each other, or preferably, on either side of a porous divider in the same tank.

Using a divider

The best divider is one which allows them to see and smell each other: plastic 'egg crate' light diffusing grating, available from most improvement stores, fits the bill here. The 1cm (½in) lattice allows the fish to see each other and permits free exchange of water between the two compartments making filtration

Above: *Target fish are useful in helping to cement pair bonds. Here two pairs of Heckel's Threadfinned Acara,* Acarichthys heckelii, *square-off over a territorial marker.*

and heating easier and allowing the fish to sense each other chemically. With good feeding, and if they are indeed of opposite sex, the intended consorts will often begin displaying to each other across the divider. At this time the divider should be removed and the two fish allowed to encounter each other. Of course, the wise aquarist will take the time to watch that encounter. Often the 'pair' must be separated right away if one or the other is not ready, and it may take many weeks and many tries for them to 'get it right'. The behavioural correlates of successful pair bonding usually involves a series of lateral displays where both fish approach each other, often with gills puffed out, and orient head to tail, side by side, often 'beating' the water between them. Genital butting and jaw-locking are also components of courtship behaviour. And if one of them 'blinks' – pulls back from the courtship retuals – the tryst is over: courtship quickly turns to aggression and the vanquished risks being liquidated if not removed from the tank speedily.

Breeding from incompatible fish
Sometimes, cichlids never get it right, even if they are male and female. This is often the case with large, wild cichlids. All is not lost, however. Incompatible fish can be spawned using the 'incomplete divider' method. This works particularly well if egg-crate divider material is available. A suitable substrate, usually a flat stone, is placed on the bottom, under and straddling the divider. Ripe females will often lay eggs on this stone and the male can fertilize them from his side of the divider, particularly if the filter outputs are set up to provide a current that favours broadcast of the milt or sperm in the female's direction. Usually, considerably less than one hundred per cent fertilization will be achieved, but for many of the larger cichlids whose spawns can approach 1000 in number, this is actually a blessing! Upon hatching, both parents will care for the fry, and it is wonderful to watch them swimming back and forth from parent to parent across the aquarium divider.

Another twist on this method takes advantage of the fact that many female neotropical cichlids are considerably smaller than their male consorts. By cutting a series of holes in the divider, large enough to allow the female to pass, but too small to permit the male to follow, the female can determine her own access to the male. She simply swims to his side of the tank when she is ripe and ready. This has been called the 'hidey-hole' method.

Conditioning breeders
Assuming a compatible pair, or a mating group of appropriate sex ratio, the fish must be conditioned by the feeding of a variety of rich, often frozen and live foods. The fish should be fed regularly and often, and appropriate water quality should be maintained. It may take several weeks, or longer, to bring wild fish into a proper condition of ripeness.

Successful systems should provide both shelter and appropriate spawn receptacles. For cave spawners, like many of the dwarf cichlids and the pike cichlids, inverted ceramic flower pots with a notched rim or with the drainage hole enlarged should be provided. For polygynous species, one per female is required. Smooth rocks or slate are often the choice of substrate-spawning cichlids. Even waterlogged leaves, dried oak or rubber tree may be useful for Acaras (eg the *Bujurquina* sp.) that practise 'moveable platform spawning'. It is often useful to introduce the female first so that she may become familiar with the environment and stake out territory before dealing with the larger, ardent male. If target fish are used, make sure they have hiding places: PVC piping of appropriate diameter and cut to appropriate lengths provides useful place of escape.

The actual spawning may often be triggered by a combination of water and temperature changes. Typically, elevated temperatures will do the trick. Slowly raising the water temperature 2-3°C (4-6°F) over a period of hours or days, even as high as 32°C (90°F) for some species (be careful – watch the fish for signs of distress and always have lots of aeration) will often trigger spawning. Combine

Right: *Some eartheaters and Acaras, like this female Parana Port cichlid,* Cichlasoma paranense, *choose platforms, like this to hold their spawns.*

this with partial water changes (20-50 per cent) using un-aged tapwater of the same temperature. For some species, the addition of 'pure', RO processed water in modest amounts will simulate the rainy season. Again, easy does it – go slowly in changing all of these parameters and be patient! It often takes weeks of cyclical changes to stimulate a spawning. Also, throughout all your labours you have to always keep in mind that some species may never spawn in captivity.

Above: *This female dwarf cichlids,* Apistogramma agassizii*, in broodcare coloration, vigorously guards her newly-laid eggs.*

Rearing the fry

The cichlid aquarist with a clutch of eggs has several options. One of the joys of keeping cichlids is observing parental care. In fact, fry left with parents usually grow faster and sturdier than those removed from their care. However, parents often prove to be egg-eaters and if the species is particularly rare, intervention may be appropriate. Frequently egg-bearing vanishes after a few spawns and is associated with young, inexperienced or transiently infertile pairs. Sometimes, traffic by their tank will cause a shy pair to terminate care: try placing dark paper over the front glass with a small peephole to watch through. If egg- or fry-eating continues beyond the first few spawnings, elect to play foster parent. The egg-bearing substrate, assuming it is removable, may be pulled from the tank and placed in a smaller tank containing clean water siphoned from the main aquarium. Alternatively, the parents may be removed. Gentle aeration will ensure a steady supply of oxygenated water for the eggs and will keep debris from settling on them. Of course, an ideal water temperature should be maintained at all times.

Bacteriacides and fungicides

Some aquarists advocate the addition of chemical bacteriacides or fungicides like methylene blue or acriflavine to the water. Others prefer transfer of the eggs to fresh tapwater, insisting that the chlorine (or chloramine) acts as a disinfectant. In fact, you cannot hatch an infertile egg; however, these chemicals can keep the spread of harmful microorganisms from fungusing 'good' eggs to a minimum. These dyes are available commercially and aquarists are advised to consult other hobby books on their usage. Typically half the usual medicinal dose is recommended.

Another alternative is to remove the fry to another tank. After about four days, the eggs hatch and the larvae, called 'wrigglers' because they can not yet swim but still flex their tails rapidly in a wriggling motion, are maintained in a ball by the parents. Some wrigglers have sticky secretions produced by special glands on their heads that allow them to be attached to plants by their parents who spit them there. Parents will often excavate a series of pits and move the wrigglers, carrying them in their mouths, from pit to pit. They eventually exhaust all of their yolk, develop fins and a workable mouth and become free-swimming and capable of feeding about four days later. The actual schedule is dependent both on the species in question and the temperature of the water – the warmer the water the faster the development.

Wrigglers are quite delicate, but free-swimming fry may be carefully siphoned out with a large-diameter siphon tube. It is often a good idea to leave a few fry with the parents. Often, when a spawn is snatched, the parents will quarrel, sometimes fatally. The fry must then be fed several times a day. Fry too small to take newly-hatched brine shrimp (*Artemia nauplii*) may be started off on commercial liquid fry food, microworms, or encapsulated rotifers (available commercially) until such times as they can handle *Artemia nauplii*. These are hatched as per instructions in most hobby texts or on the container of eggs, which are available at most shops. Certain fish will accept frozen newly-hatched brine shrimp, or even crumbled dry food, but growth rates and general health are much more satisfactory with live first foods.

The fry tanks should, of course, be filtered. Sponge filters from established tanks are preferred. Regular partial water changes, small but frequent, are recommended, as are the addition of snails or regular siphoning of the bottom to dispose of the uneaten and decaying shrimp.

Fry left with parents will be tended for several weeks. Typically both male and female will provide perimeter defence and herd the fry around the tank as they forage. It is interesting to watch the parents 'call' their youngsters with a flick of the ventral fins when danger approaches: they all drop to the bottom beneath the parents where they may be collected and moved if need be. Fry left with parents must be fed, as explained above, when free-swimming. It is a good idea to remove other tank occupants at this time, or risk loss of the spawn. In particular, predatory catfish will dispatch a clutch in no time after the lights go down. One solution is to leave the lights on 24 hours a day until the fry are large enough to escape.

Below: *Many eartheaters, like this Red Hump female,* 'Geophagus' steindachneri, *are mouthbrooders and continue to offer their fry buccal protection.*

In the case of mouthbrooding cichlids, like the *Bujurquina* species or many of the Eartheaters, fry-tending includes uptake of the young into the buccal (throat) cavity. Some cichlids, like Uaru and Discus, actually provide initial nutriment in the form of mucoid secretions on their flanks which the fry feed on (contact feeding). Of course, custodial care has its temporal limitations, typically 6-8 weeks for most species, although some pike cichlids have been reported to defend 'fry' for as long as one year. Tired parents make their condition known and actually appear to avoid their youngsters. Failure to remove the fry at this point usually results in them being eaten by the parents, often as a prelude to the next spawning. Once on their own, juveniles respond positively to the conditions outlined for adults. At this point, they can be shared with aquarist friends or sold to shops or wholesalers to defray the cost of their upkeep.

Breeding aquarium fish, and cichlids in particular, is a wonderful extension to the tropical fish hobby, accessible to all aquarists with the interest and energy to provide the appropriate conditions. You do not need many tanks and a lot of space: even the smallest of the dwarf cichlids provide the full range of cichlid parental behaviour. When cared for properly, you cannot stop fish from spawning, and that is the ultimate sign that the captive environment you have created is successful!

The enquiring aquarist will find the following species sections an easy-to-consult reference source, providing essential data on the most popular cichlids in fishkeeping today. Dr Paul Loiselle's African section is surely beyond improvement and is a wonderful example of how knowledge, enthusiasm and complete confidence can illuminate when an author knows the subject from first-hand experience. Dr Loiselle led the advance of the new age of cichlid writers and, whilst having the authority to use scientific terms precisely, he does not appear to forget to provide fundamental aquarium information. Cichlid enthusiasts need to locate basic data quickly, especially information related to maintaining and ultimately spawning species in their care. Dr Wayne Leibel's approach to South American cichlids illustrates his up-to-date knowledge on the changing nomenclature. Dr Leibel's speciality is the cichlids of South America and his enthusiasm is famous

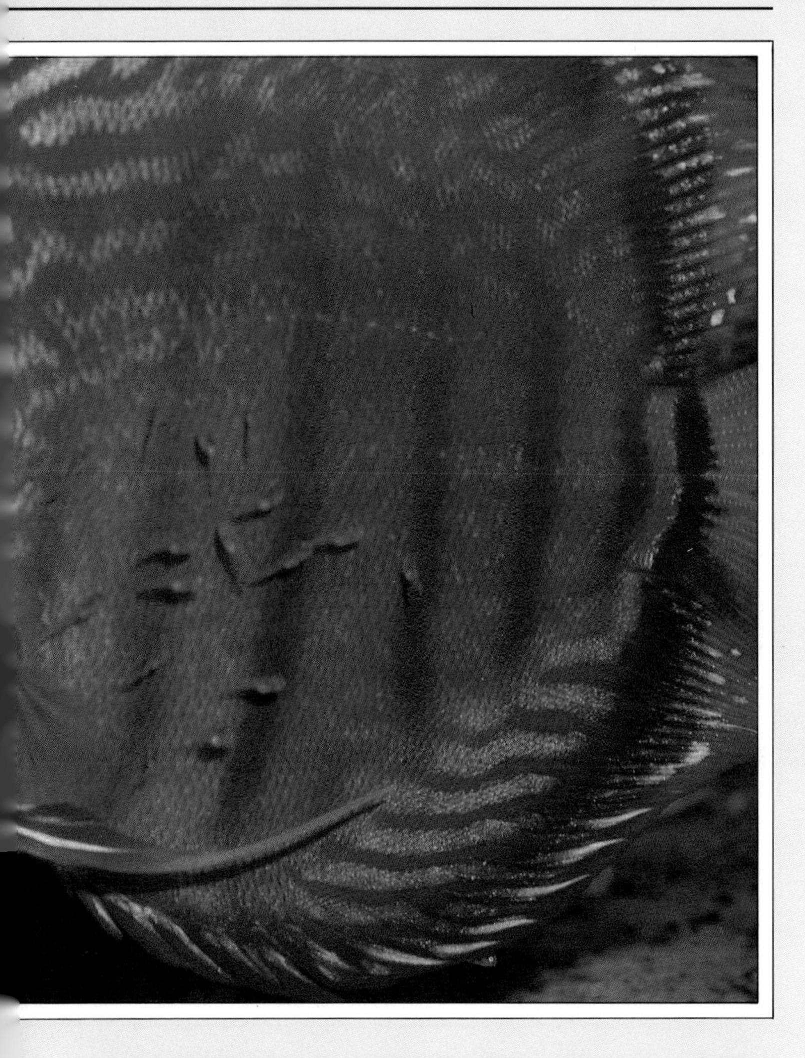

amongst hobbyists across America. This continent is home not only to the 'King of the Amazon', Discus, but also to the best known archetypal aquarium fish – the angel. It is hard to think of the hobby without the multitude of colour forms available today.

My own contribution to the sections dealing with Central American cichlids is mostly based on my early import experiences gained as a great number of new species came into the European hobby throughout the 1980s. It seemed that every week a new *'Cichlasoma'* would arrive on the scene and solicit fresh excitement for enthusiasts. Probably a case of *déja vu* for Dr Loiselle who saw African cichlids go through a similar awakening in the 1970s. After writing the Central American cichlid book I spent many happy months with renowned cichlid importer Don Conkel in Florida. Much of what I learned confirmed that the hobby is surely blessed by such a fantastic group of fish.

African cichlid species

This section features a representative selection of African cichlids. Fish are grouped by region of origin and introduced by a summary of the common features that influence their care in captivity. Details of the natural habitat, standard length, sexing, feeding type, compatibility and breeding behaviour of individual species follow. The first group features the riverine cichlids and includes those from Lake Victoria. Here are the jewel fish, the *Chromidotilapia* lineage, tilapias and, of course, the haplochromines. They, together with the tilapias, are native to the rivers of eastern and southern Africa and dominate all the northern Great Lakes, other than Lake Tanganyika. In Lake Tanganyika, the second region to be considered, five distinct evolutionary lines of cichlids are represented. Here the lamprologines predominate, followed by the featherfins (*Ophthalmochromis*), with haplochromines a poor third. The largest and smallest known cichlid

species are native to Lake Tanganyika, and cichlids occur in every habitat within the lake and dominate many of them. Lake Malawi, the third region, is home to some 250 described cichlids, almost all haplochromines. It is considerably older than Lake Tanganyika and many of its cichlids have evolved a specialized life style, as exemplified by the rock-dwelling mbuna.

Research continues into the evolutionary relationships of African cichlids and recent revisions propose new generic names for cichlids included in the genera *Haplochromis* and *Lamprologus*. We have chosen to use the generic names proposed for the Malawian species formerly included in the genus *Haplochromis* and to treat the new names proposed for the extra-Malawian representatives as subgenera. However, we have not incorporated revisions that would split *Lamprologus* into smaller genera.

RIVERINE AND LAKE VICTORIA CICHLIDS

Jewel fish and their allies

The dwarf to medium-sized monogamous jewel fish are unique among African cichlids in dividing long-term responsibility for their fry equally between the sexes. All do well in moderately hard (6-10°dH), neutral to slightly alkaline water and most thrive at temperatures of 21-27°C (70-80°F), increased to 30°C (86°F) for breeding. Although they tolerate short-term exposure to nitrite better than many west African cichlids, this is no excuse for sloppy nitrogen cycle management. Like other hardy cichlids, jewel fish and their allies look better – and breed more readily – under a regime of effective biological filtration and frequent partial water changes. These cichlids may dig in conjunction with spawning, but do not pose a serious threat to well-rooted plants. They all appreciate a layer of floating plants. In a sufficiently large tank jewel fish are easy to breed; at 27°C (80°F) eggs hatch after 40-72 hours and fry are mobile four days later. Offer them brineshrimp nauplii and finely powdered prepared foods. Remove fry from their parents four to six weeks after spawning, otherwise they may be eaten. When culling spawns or selecting fish for breeding, remember that males grow faster than females.

Anomalochromis thomasi

Dwarf jewel fish

● **Distribution:** Coastal rivers of Guinea, Sierra Leone and Liberia.

● **Habitat:** Clear and blackwater streams flowing under intact or recently disturbed rain forest.

● **Length:** Males up to 7.5cm (3in); females up to 5.7cm (2.25in).

● **Sexing:** Females often show a pattern of red dots in the shoulder region, as well as a pattern of bars on their flanks when courting. Males have longer soft dorsal and anal fins.

● **Diet:** Omnivore. Offer live food to encourage breeding.

● **Compatibility:** An excellent general community resident that poses no risk, even to small schooling fish. Pairs defend territories 60cm (24in) square, so it is possible to house several pairs in a well-planted tank 90-120cm (36-48in) long. May prove shy without floating plants.

Below: **Anomalochromis thomasi** *An ideal cichlid for beginners.*

Above: **Hemichromis elongatus** ♀ *A territorial female.*

Raising a group of youngsters to maturity together is the easiest way to secure a compatible pair. Spawns readily in a community tank, producing up to 500 eggs, but pairs seldom succeed in raising young to independence in the presence of fast-swimming schooling fish and scavengers. Prefers to spawn in a partially enclosed space and sometimes deposits its eggs on a large plant leaf. Hatching occurs in 40-48 hours at 27°C (80°F). The fry begin swimming about three days later, but do not absorb the remainder of their yolk-sacs for another 12-16 hours. Offer them microworms for the first 3-4 days, by which time they will have grown sufficiently to feed on brineshrimp nauplii and finely powdered prepared food. It is not unusual for young pairs to eat their first few spawns before settling down to model parenthood. The fry grow rather slowly, reaching 1cm (0.4in) after one month, even with good feeding and frequent partial water changes. They are more sensitive than other jewel fish to nitrogen cycle mismanagement. Sibling cannibalism is not a problem in the dwarf jewel fish. Brood care persists for about four weeks after hatching and the young become sexually mature after 8-10 months, at 4.5cm (1.8in).

A slow-growing, but long-lived, hardy and easily bred dwarf cichlid. Aquarium strains are descended from the typical Sierra Leonian populations of *A. thomasi*. Liberian fish differ in coloration and may represent a distinct species.

Hemichromis elongatus

Five-spot jewel fish

● **Distribution:** Coastal rivers in Guinea, Sierra Leone and southwestern Liberia; Togo southwards to northern Angola; lower Niger basin; Zaire (Congo) basin, including Shaba (Katanga) region; upper Zambezi basin.
● **Habitat:** Small streams, shallows of large rivers, marshes and oxbow lakes. Usually found close to submerged wood or aquatic plants. In the southern portion of its range, it penetrates coastal lagoons where the salinity can approach that of sea water.
● **Length:** Males up to 15cm (6in); females up to 10cm (4in).
● **Sexing:** Females are smaller and fuller bodied.
● **Diet:** Piscivore.
● **Compatibility:** Aggressive and violently intolerant of its own species. Devours any companion small enough to eat. Keep only one pair in tanks less that 180cm (72in) long. House with tilapias or large neotropical cichlids of a like disposition.

In a community setting the female is usually safe from injury by the male, but tankmates risk serious injury or death unless they can move out of the pair's territory. The pair bond tends to be unstable in the absence of target fish. The 'incomplete divider' method is suitable for breeding isolated pairs. In spawning sites close to cover they may produce up to 1000 eggs. The fast-growing fry are given to sibling cannibalism. They are sexually mature at about eight months at 7.5cm (3in).

Above: **Hemichromis guttatus** ♂ *The female is more intensely red.*

Hemichromis guttatus

Common jewel fish

● **Distribution:** Coastal rivers in Ivory Coast and Ghana west of the Volta; western Togo to southern Cameroons.

● **Habitat:** Small streams, marshes, oxbow lakes. Usually found close to submerged wood or aquatic plants.

● **Length:** Males up to 10cm (4in); females up to 7.5cm (3in).

● **Sexing:** Females are smaller, with redder base colour and less metallic blue spangling on body and fins.

● **Diet:** Micropredator. Offer colour food regularly.

● **Compatibility:** Relatively peaceful towards companions too large to eat, except when breeding. More tolerant of its own species than *H. elongatus*. Pairs defend territories 90cm (36in) square.

Pairs readily in a large community tank, with little risk of injury to tankmates. Rearing a group of young to maturity is the easiest way of securing a compatible pair. Pair stability is reinforced by the presence of target fish. The size difference between the sexes makes the 'separate compartment' method practical for breeding isolated pairs. The fish prefer to spawn in the open on smooth rocks or similar sites. Spawns can number 500 eggs. The fry grow rapidly and are less given to sibling cannibalism than those of the five-spot jewel fish. They attain sexual maturity 6-8 months after hatching, at 5cm (2in) for males, 3.75cm (1.5in) for females.

Thysia ansorgii

Five-spot cichlid

● **Distribution:** Coastal rivers from just west of the Bandama in the Ivory Coast to the Ankombrah in southwestern Ghana, thence from the Ouémé River in Benin to the neighbourhood of Douala in the Cameroons.

● **Habitat:** Small streams flowing under intact or recently disturbed forest cover. Sometimes found in swamps and oxbows, usually in association with stands of aquatic plants.

● **Length:** Males up to 10cm (4in); females up to 7.5cm (3in).

● **Sexing:** Easily sexed. Males have long, flowing soft dorsal and anal fins, as well as an asymmetrical caudal fin with a distinct reticulate pattern. Females have a small cluster of iridescent white scales just above the vent.

● **Diet:** Omnivore.

● **Compatibility:** A 'giant' dwarf cichlid whose behaviour in captivity resembles that of *A. thomasi* in all significant respects. Introduce appropriate dither fish to overcome shyness.

Breeding details as recommended for the dwarf jewel fish (see page 104), but pairs are usually more successful in raising fry in a community setting than *A. thomasi.* This particular species prefers to spawn in a partially enclosed space and will often deposit its eggs on a vertical surface. Spawns can number up to 250 stalked eggs, which hatch in 72 hours at 27°C (80°F) and the fry are mobile 7-8 days later.

The fry are more sensitive to dissolved metabolic waste than are those of the red jewel fish, but are not difficult to rear under a routine of frequent partial water changes. The young are sexually mature 6-7 months after hatching, at 6.5cm (2.5in) for males and 5cm (2in) for females.

Below: **Thysochromis ansorgii** ♀ *Note the iridescent white ventral spots.*

The *Chromidotilapia* lineage

This large group of west African cichlids consists of the genera *Chromidotilapia*, *Pelvicachromis* and *Nanochromis*. These are dwarf to medium-sized cichlids with clear visual distinctions between the sexes. Males are larger and have longer, more elaborate finnage. Females are smaller and more colourful. A brilliantly coloured ventral spot and contrasting, often metallic, coloration in the spiny dorsal fin are typical of the female's courting dress. In all these species, the female's courtship display centres around the presentation of this vividly coloured ventral blotch to the male.

All representatives of this group are extremely sensitive to dissolved metabolites. Scrupulous attention to proper nitrogen cycle management is absolutely essential to their successful maintenance. Rapids-dwellers, such as *Nanochromis*, require high levels of dissolved oxygen to prosper and appreciate strong water movement. Ideally, therefore, combine an outside power filter with brisk aeration in their tanks. Many *Pelvicachromis* and *Nanochromis* species hail from soft, acid-water habitats, but in captivity most do well – and will even spawn – in slightly hard (6-10°dH), neutral to slightly alkaline water. However, their progeny will display a balanced sex ratio only if the pH and hardness levels in the spawning tank fall within a narrow range. A temperature range of 21-27°C (70-80°F) suffices for day-to-day management, with an increase to 30°C (86°F) to encourage breeding.

These cichlids do not eat aquatic plants, but the normal foraging behaviour of *Chromidotilapia* and the larger *Nanochromis* species poses some risk even to well-rooted specimens. Potted plants are usually immune from such disturbance. Digging is restricted to periods of sexual activity and limited to the vicinity of the spawning site. All appreciate the security of a screen of floating plants, such as *Ceratopteris cornuta*.

Chromidotilapia guntheri

Mouthbrooding krib

● **Distribution:** Coastal rivers from Ivory Coast to the southern Cameroons, including the Niger and Volta basins.

● **Habitat:** The slower stretches of streams and rivers, usually over relatively fine substrates and often close to stands of submerged wood. Less common in oxbows or lakes.

● **Length:** Males up to 18cm (7in); females up to 15cm (6in).

● **Sexing:** Females are smaller, with a rosy violet ventral blotch and a nacreous gold spiny dorsal fin.

● **Diet:** Omnivore.

● **Compatibility:** Will eat fish as large as a female guppy, but overall, an inept predator. Sexually inactive individuals ignore larger midwater-swimming species. Pairs defend territories 100cm (39in) square. They will disregard tankmakes that move beyond their territorial borders.

Rearing a group of young to maturity together should produce a compatible pair. Pair stability is reinforced by the presence of target fish. The size difference between the sexes makes the 'separate compartment' breeding approach workable with isolated pairs. *C. guntheri* is an advanced mouthbrooder that deposits its eggs on a flat surface and fertilizes them in conventional cichlid fashion. The male then takes them into his mouth. Some pairs share the incubation sequence, while in others the male is the sole custodian of the brood until, at 27°C (80°F), the mobile fry emerge, a week after spawning. Both the

Above: **Chromidotilapia guntheri** ♀ *Rosy area reddens at spawning.*

parents then share custodial duties, to the point of allowing the young to take shelter in their mouths should danger threaten. Broods rarely exceed 100 fry. The young can take brineshrimp nauplii and finely powdered prepared food for their initial meal. The fry grow rapidly and can measure nearly 2.5cm (1in) within a month of their initial release. This is the time to separate them from the adults, since they have grown too large to fit within their parents' mouths. The young attain sexual maturity at 8-10 months after hatching, at 10cm (4in) for males, 6.5cm (2.5in) for females.

This hardy, medium-sized cichlid prospers over a wide range of pH and hardness conditions and makes an excellent addition to a community of like-sized African cichlids. The mouthbrooding krib is known in the older aquarium literature as *Pelmatochromis guentheri.*

Nanochromis parilius

Nudiceps

● **Distribution:** The lower Zaire River, from the Malembo (Stanley) Pool to its mouth.
● **Habitat:** Rippling pools of small streams flowing into the main channel of the Zaire River, as well as calm stretches of the main river itself, typically over coarse gravel.
● **Length:** Males up to 7.5cm (3in); females up to 5.7cm (2.25in).
● **Sexing:** Females are much deeper bodied, with shorter soft dorsal and anal fins.
● **Diet:** Omnivore. Live foods encourage breeding.
● **Compatibility:** Males are very intolerant of other males of their own species. Even sexually inactive individuals of both sexes often behave aggressively towards other small bottom-dwelling fish, but will ignore midwater-swimming tankmates too large to make a comfortable mouthful. Best housed in tanks at least 90cm (36in) long, as males are often very aggressive towards sexually unreceptive females.

Even sexually quiescent females appear ripe, for a conspicuous ovipositor is visible at all times. However, truly ripe individuals appear ready to burst. This species is best maintained on a harem basis to reduce the risk of injury or death to the female during courtship. If only a single pair is available, it is essential to provide an abundance of hiding places and suitable target fish in the breeding tank. Courtship is protracted and it is not uncommon for a young

Above: **Nanochromis parilius** ♀ *Has a rounder ventral profile.*

female to eat her first few spawns. As with the common krib, the initial phases of brood care are exclusively maternal and the female's refusal to allow the male to enter the breeding cave is a reliable indication that spawning has occurred. Spawns can number up to 100 eggs, but are usually smaller. Hatching occurs in about 48 hours at 27°C (80°F) and the female brings the free-swimming fry out of the cave five days later. At this point, the male may join his mate in fry care. The robust young can take brineshrimp nauplii and finely powdered prepared food for their initial meal. The fry are very sensitive to high ambient bacterial and dissolved metabolic waste levels.

Twice-weekly partial water changes of 50 percent during the first month help to avoid losses and maximize growth. Parental care lasts about four weeks in captivity. The fry are sexually mature 6-8 months after spawning, at 6cm (2.4in) for males, 4cm (1.6in) for females.

Pelvicachromis pulcher

Kribensis; Common krib

● **Distribution:** Coastal rivers of southern Nigeria, from the Niger delta westward to the Benin border.

● **Habitat:** Small streams and the slower stretches of rivers, usually over relatively fine substrates and always closely associated with stands of aquatic plants.

● **Length:** Males up to 10cm (4in); females up to 7.5cm (3in).

● **Sexing:** Females are smaller and deeper bodied, with a bright red ventral blotch and a metallic gold spiny dorsal fin.

● **Diet:** Micropredator. Live foods encourage breeding.

● **Compatibility:** An excellent general community resident, but may prove shy unless kept in a well-planted tank with small schooling fish. Pairs defend territories 60cm (24in) square, so it is possible to house several pairs in a tank 90-120cm (36-48in long). Restricts digging to spawning site.

A single male and female will pair readily in a community setting, but rearing a group of young to maturity remains the surest route to a compatible pair. Pair stability and parental reliability are reinforced by the presence of target fish. Courtship is protracted and pairs often go through several false spawns before finally producing a clutch of eggs. Young pairs may eat their first few spawns, but eventually settle down to model parenthood. Up to 100 eggs are placed on the roof and sides of an enclosed site and tended exclusively by the female. Her refusal to allow the male to enter the spawning site is a clear indication that spawning has occurred. The eggs hatch in 48-50 hours at 27°C (80°F) and the female brings the free-swimming fry out of the cave five days later. The young can take brineshrimp nauplii and finely powdered prepared food for their first meal. Rearing them poses no problems if they are kept under a routine of frequent partial-water changes. Parental care lasts from 4-6 weeks in captivity. A pH of 6.8-7.2 will result in a balanced sex ratio among the young, which can be reliably sexed on the basis of differences in ventral fin shape 4-5 months after spawning. They attain sexual maturity three months later, at 6.5cm (2.5in) for males, 4.5cm (1.8in) for females.

Pelvicachromis sacrimontis apparently ranges from the Niger Delta to the Cross River, replacing *Pv. pulcher* in eastern Nigeria. It is imported and marketed under the name of Giant Krib, and is immediately recognizable by the absence of metallic gold coloration in the spiny dorsal fin of either sex. Maintenance requirements as for *Pv. pulcher*, but a pH of 5.5-6.0 is required to produce a balanced sex ratio among the fry.

Below: **Pelvicachromis pulcher** *Fin markings vary in males.*

Tilapias and their allies

Representatives of this economically important cichlid lineage occur throughout Africa and even into the Near East. A few *Oreochromis* species are endemic to one or other of the Great Lakes, and Barombi Mbo, a crater lake in the Cameroons, supports a unique collection of tilapia-descended genera. However, the majority are river dwellers, so it seems appropriate to consider them under that heading. Most tilapias are medium-sized to quite large cichlids, although a few dwarf species are known. As juveniles, they all show the typical black 'eye-spot' or 'tilapia-spot' at the base of the soft dorsal fin. This spot disappears from the colour pattern of mouthbrooding representatives of the group, but usually persists into adulthood among the substrate spawners.

Despite their hardiness and attractive coloration, tilapias do not enjoy a great deal of popularity among aquarists. Most tilapias are omnivores with strong herbivorous tendencies, and their habit of treating planted aquariums as self-service salad bars is not appreciated by most aquarists.

Steatocranus and *Chaytoria* species are sensitive to dissolved metabolites and require the same attention to proper nitrogen cycle management as representatives of the *Chromidotilapia* lineage. The remaining tilapias can tolerate elevated nitrite levels for considerable periods of time without suffering ill-effects although, like all hardy fish, they look their best – and are more likely to spawn regularly – under a regime of frequent partial-water changes. Tilapias as a group are indifferent to the chemical make-up of their water as long as extremes of hardness or pH are avoided. Most *Tilapia*, *Sarotherodon* and *Oreochromis* species appreciate a temperature range of 21-27°C (70-80°F) for day-to-day maintenance, with an increase to 30°C (86°F) for breeding.

Oreochromis mossambicus

Mozambique mouthbrooder

● **Distribution:** Coastal basins from Quelimane in southern Mozambique southward to the Pongola River in South Africa. South of this point, this species occurs in brackish water habitats as far as Algoa Bay.

● **Habitat:** The lower reaches of rivers in areas of moderate current, as well as in coastal lagoons, oxbow lakes and swamps. Typically found over soft bottoms, often in well-planted habitats.

● **Length:** Males up to 38cm (15in); females up to 25cm (10in).

● **Sexing:** Females are smaller than males and lack their bright colours and excavated cranial profiles.

● **Diet:** Omnivore with strong herbivorous tendency.

● **Compatibility:** Males defend a territory centred on a huge spawning pit with a diameter roughly twice their own overall length. Adult males are aggressive towards males of the same species and other cichlids with a similar colour pattern. Single individuals make good cichlid community residents if kept with tankmates of a like size and disposition. Although it will eat fish small enough to be easily swallowed, *O. mossambicus* is a clumsy predator.

Short of physically separating the sexes, there is no way to prevent this species from breeding in captivity. Males attempt to lure ripe females into their spawning pit with a vigorous 'headstand' display. The two fish then begin a reciprocal circling pattern that leads to the expression of a batch of eggs by the female. She immediately picks them up and nibbles at the male's vent to elicit

Above: **Oreochromis mossambicus** ♂ *Male digging a nest pit.*

ejaculation. The eggs are thus fertilized in her mouth. Once the female is spent, the male chases her from his territory and attempts to repeat the performance with another female. The incubation period lasts 14 days at 27°C (80°F) and, in nature, the female will continue to protect the mobile fry for an additional week. A large female can produce up to 1700 eggs, but spawns of 150 to 200 fry are the usual rule in captivity. The robust fry can take fine prepared foods and brineshrimp nauplii for their first meal. They are easily reared, with the males growing faster and bigger than females and both sexes reaching sexual maturity relatively easily. With heavy feeding and frequent water changes, it is not unusual for the young to begin breeding when only six months old, at a length of 5.75cm (2.25in) for males, 4.5cm (1.8in) for females.

Steatocranus casuarius

Bumphead/Buffalohead cichlid

● **Distribution:** The lower Zaire River, from the Malembo (Stanley) Pool to Matadi.

● **Habitat:** Stretches of relatively calm water between rapids in the main channel of the Zaire River, typically over rocks.

● **Length:** Males grow up to 16.5cm (6.5in); females up to 10cm (4in).

● **Sexing:** Females are smaller and lack the male's prominent nuchal hump and long, flowing soft dorsal and anal fins.

● **Diet:** Herbivore.

● **Compatibility:** Somewhat aggressive towards its own species and other bottom-dwelling cichlids. It is possible to keep more than a single pair in a large aquarium 120cm (48in) long, generously furnished with caves and similar hiding places. They are apt to be shy unless housed with suitable target/dither fish, which are ignored outside the particular periods of reproductive activity. These cichlids move large volumes of gravel when preparing a spawning site, but despite this behaviour will not usually disturb potted plants.

A monogamous, cave-spawning cichlid that pairs and spawns readily in a community setting. Courtship is not obvious. A sudden increase in digging activity followed by the appearance of a short white ovipositor about 24 hours before spawning are the most reliable indications that reproduction is in the offing. Up to 100 large, pear-shaped eggs are placed on the roof and sides of a cave. The division of parental roles is comparable to that in the common krib (see page 110). The eggs hatch after five days at 27°C (80°F), and the robust fry emerge from the cave 4-5 days later. Treat them as recommended for the fry of *N. parilius* (see page 109). Adults are extremely protective of their fry and usually have no difficulty rearing them to independence in a community tank, although this approach to breeding can be rather hard on its other residents. Parental care persists until the female ripens a new clutch of eggs, 6-8 weeks after spawning. If the breeding tank is large enough, there is no need to remove the fry at this point. They do not prey on their younger siblings and, in consequence, the fry are tolerated within the breeding territory by their parents. The young are considered sexually mature about six months after spawning, at 5cm (2in) for males, slightly less for females.

Above: **Steatocranus casuarius**
♂ Has prominent nuchal hump.

Below: *♂ Only the male sports long soft dorsal and anal fins.*

Above: **Tilapia mariae** ♀ *Clearly showing five spots.*

Tilapia mariae

Five-spot tilapia

● **Distribution:** From the Bandama River in Ivory Coast to the Bia River in southwestern Ghana; thence from Lake Toho in Benin to the Kribi River in the southern Cameroons.

● **Habitat:** The lower reaches of rivers flowing under forest cover, as well as in coastal swamps and lagoons. Abundant in deep, flowing pools where submerged wood, rocks and stands of algae-coated aquatic plants are plentiful.

● **Length:** Males up to 30cm (12in); females slightly smaller.

● **Sexing:** Not easily sexed. Males grow slightly larger than females, while the red shoulder spot of the female is usually brighter. Large females often have a red 'bib' on the chest immediately in front of the ventral fins. Males are more aggressive towards their companions, whereas females tend to behave submissively.

● **Diet:** Omnivore with strong herbivorous tendencies.

● **Compatibility:** Adults are extremely aggressive towards their own species. Single individuals make good cichlid community residents if kept with tankmates of a like size and disposition. House only one pair per tank. They may make life extremely unpleasant for their companions, regardless of species, with the onset of reproductive behaviour.

Handle this monogamous species in the same manner as the common jewel fish (see page 106). Breeding efforts are most likely to succeed with young pairs in a tank 120cm (48in) long. Spawns numbering as many as 5000 eggs are often placed within a partially enclosed space or on a vertical surface. The female is the primary caretaker of the developing zygotes, while the male defends the pair's territory against intruders. The eggs hatch in 60-72 hours at 27°C (80°F). The female may move the fry several times to pre-dug pits before they become mobile three days later. Both parents vigorously defend the free-swimming young. The fry are easily raised on a diet of brineshrimp nauplii and powdered prepared foods. With frequent water changes, they reach 4cm (1.6in) after 12 weeks. Parental care can persist at least this long in captivity, but most aquarists prefer to separate the fry from their parents by the end of the second month after spawning. Like the young of most tilapias, those of *T. mariae* are sexually precocious. Reproductive activity begins 6-8 months after spawning, at 6.5cm (2.6in).

Haplochromis and their allies

The great majority of these cichlids are small to medium-sized fish. All are openly polygamous, maternal mouthbrooders, in which the males grow faster and bigger than females and both sexes reach sexual maturity relatively early. In most species the sexes are easily distinguished; males are brilliantly coloured and females are drabber. In the great majority of *Haplochromis* species, the male sports conspicuous pale yellow to orange spots on the anal fin about the same size as that species' eggs. During spawning, the male displays these pseudo-ocelli conspicuously to the female, who attempts to take them into her mouth, just as she would her unfertilized eggs. This mouthing behaviour triggers ejaculation, so that in trying to pick up the 'egg dummies' she collects a quantity of sperm that fertilizes the eggs already in her mouth. *Pseudocrenilabrus* and some species native to the rapids of the Zaire basin lack such pseudo-ocelli. Their spawning behaviour resembles that of *Oreochromis mossambicus*, with the egg-laying female nibbling directly at the male's vent.

These cichlids are essentially indifferent to water chemistry, but dislike extremes of pH and hardness. Specialized rapids-dwelling species and those native to the northern Great Lakes are sensitive to dissolved metabolites and require regular partial-water changes. Haplochromines from quieter river waters and isolated shore pools of large lakes can tolerate elevated nitrite levels for considerable periods, although such hardiness is no excuse for inadequate nitrogen cycle management. Most extra-Malawian haplochromines tolerate temperatures as high as 35°C (95°F) if their tank is well aerated. Although the incubation period is shorter in warmer water, males are also considerably more aggressive at higher temperatures. For daily maintenance, temperatures of 21-24°C (70-75°F) will suffice, with an increase to 27-30°C (80-86°F) to encourage breeding.

It is unfortunate that these haplochromines have been somewhat overshadowed by their distant Malawian cousins. They are just as colourful, will thrive and breed in smaller tanks and all are indifferent to rooted aquatic plants. Not always available, but worth taking some trouble to find.

Haplochromis (Astatotilapia) burtoni

Burton's mouthbrooder

● **Distribution:** The basin of Lake Tanganyika, including the Malagarasi River and associated swamps.

● **Habitat:** Small streams, swamps and isolated shore pools along the coast of Lake Tanganyika. Uncommon in the lake itself, which seems to serve chiefly as a dispersal corridor to preferred peripheral habitats.

● **Length:** Males up to 10cm (4in) and females up to 7.5cm (3in) in captivity; 6.5cm (2.5in) and 5cm (2in) in nature.

● **Sexing:** Males are brilliantly coloured; females are drabber.

● **Diet:** Omnivore. Offer colour food regularly.

● **Compatibility:** Males are very aggressive towards one another and towards other male haplochromines with a similar colour pattern. Very large males are too bellicose to be kept together in tanks less than 150cm (60in) long. Groups of several younger males can be maintained successfully in tanks more than 120cm (48in) long. Midwater-swimming fishes too large to make a convenient mouthful are ignored. Since males move extensive areas of gravel during the construction of a nest pit, potted plants are less likely to be disrupted than rooted ones.

Above: **Haplochromis ishmaeli** ♂ *Male has conspicuous 'eggspots'.*

H. burtoni is one of the easiest African cichlids to breed. Spawns can number up to 120 eggs, but 50 is closer to the norm. The incubation period lasts 10 days at 30°C (86°F). The fry are large enough to take brineshrimp nauplii and finely powdered prepared foods for their initial meal. They are easy to rear when handled in the same fashion as those of *Oreochromis mossambicus* (see page 112). This species matures remarkably quickly; with frequent feedings and water changes, the young can attain sexual maturity eight weeks after release, at a length of 3.75cm (1.5in) for males, 2.5cm (1in) for females.

Haplochromis (Prognathochromis) perrieri

Green arrow haplochromis

● **Distribution:** Lake Victoria.
● **Habitat:** Vegetated inshore habitats throughout the lake.
● **Length:** Males up to 10cm (4in); females slightly less.
● **Sexing:** Males are brilliantly coloured; females are drabber.
● **Diet:** Piscivore.
● **Compatibility:** Males behave aggressively towards one another. Several individuals will coexist in a tank more than 150cm (60in) long, but only one will develop the typical courting dress of the species and enjoy reproductive success. Females can also be somewhat snappy towards one another if crowded, and individuals of both sexes may react unfavourably to other haplochromines with the same overall appearance. Will prey efficiently upon fish as large as a female guppy, but ignores larger midwater-swimming companions. May excavate a nest pit, but not as energetically as *H. burtoni.*
● **Breeding:** Do not attempt to maintain a breeding group in a tank less than 120cm (48in) long. Apart from its need for more swimming room, *H. perrieri* requires the same

Below: **Haplochromis pellegrini** ♀ *A ripe female.*

management as other small, openly polygamous cichlids. Spawns can number up to 75 eggs, but 35 is more usual. The incubation period lasts 14 days at 30°C (86°F). Offer the fry brineshrimp nauplii and finely powdered prepared foods for their first meal. Given scrupulous attention to cleanliness, they are easy to rear and can be reliably sexed on the basis of colour differences 12 weeks after release. However, reproductive activity does not commence until the young are at least six months old, at a length of 7cm (2.75in) for males, slightly less for females.

Pseudocrenilabrus multicolor

Egyptian mouthbrooder

● **Distribution:** The Nile River drainage from Lake Albert and the Murchison Falls northwards to the Delta.

● **Habitat:** Swamps, small streams and the peripheral waters of rivers and lakes, usually in heavily planted habitats.

● **Length:** Males 6.5cm (2.5in); females 5cm (2in).

● **Sexing:** Females are drabber.

● **Diet:** Omnivore. Provide live food and colour food.

● **Compatibility:** Males are very aggressive towards one another at all times and, with the onset of breeding, prove surprisingly belligerent towards other bottom-dwelling fish. Breeding territories measure about 45cm (18in) square, which are very large for such a small fish. Usually does poorly when housed with larger *Haplochromis* species, but does surprisingly well in the company of *Oreochromis* species. Ignores midwater-swimming companions too large to be easily swallowed.

● **Breeding:** The Egyptian mouthbrooder is very easily bred. As males court vigorously, this species is best housed in tanks more than 90cm (36in) long. Spawns can number 100 eggs, but 35-50 is closer to the norm. Incubation lasts 10 days at 30°C (86°F). Young females may fail to carry their initial brood to term, but can be counted upon to complete subsequent efforts successfully. Raise the fry on a diet of brineshrimp nauplii and powdered prepared foods. With frequent water changes, they can attain sexual maturity four months after spawning, at a length of 2.5cm (1in) for males, slightly less for females.

Below: **Pseudocrenilabrus multicolor** ♂ ♀ *The male is shown above.*

LAKE TANGANYIKA CICHLIDS

Despite the wide diversity of size, form and reproductive patterns among the Lake Tanganyikan cichlids, they share sufficient features to warrant a common approach to their care in captivity. All require hard, alkaline water to prosper and all are extremely sensitive to abrupt changes in both the chemical composition and temperature of their water. Most species prefer temperatures in the range of 22-24°C (72-75°F) for day-to-day maintenance, with an increase to 27-30°C (80-86°C) for breeding. Without exception, these cichlids are highly susceptible both to ammonia and nitrite poisoning. They dislike large-scale water changes, so the fishkeeper must rely on light stocking rates, efficient biological filtration and small (10-15 per cent), frequent water changes for effective nitrogen cycle management.

Lamprologus and their allies

Some 48 species of *Lamprologus* from Lake Tanganyika have been described and exported as ornamental fish, and all but the giant predator *Boulengerochromis microlepis* have been well received by aquarists. *Cyphotilapia* and the larger *Lamprologus* species are formidable piscivores, whereas *Telmatochromis* and the medium-sized *Lamprologus* are micropredators. *Chalinochromis* and *Julidochromis* employ their specialized, tweezer-like teeth to pluck invertebrates from the algal and sponge mats that encrust the lake's rocky substrate. Their diet in captivity should reflect this reliance on animal food. These cichlids dig when breeding, but restrict their excavations to the immediate vicinity of the spawning site and pose little threat to rooted plants, such as Java or African fern.

Chalinochromis, *Julidochromis* and most *Lamprologus* and *Telmatochromis* species are monogamous in nature. Most of the smaller ostracophil, or shell-dwelling, *Lamprologus* and *Telmatochromis* species are harem polygynists. However, many species known to practise monogamy in the wild tend to shift to a polygynous system in captivity if a surplus of females is present. As a rule, pair (or harem) bonds are very strong in lamprologines, so it is possible to maintain breeding pairs or groups in isolation with minimal risk to the female. Courtship is not usually overt and pairs are very secretive about spawning. The female's reluctance to leave the cave usually indicates that spawning has occurred, but often the first indication is the appearance of free-swimming fry! Eggs hatch in 40-48 hours at 27°C (80°F); the fry are mobile five days later.

Save for *Cyphotilapia*, a maternal mouthbrooder, these substrate-spawning cichlids prefer enclosed spawning sites. Both parents share responsibility for fry, much like the *Pelvicachromis* species. However, the young are protected only while they remain within their parents' territory and no effort is made to retrieve any that wander further away. In other respects, lamprologine reproductive behaviour displays great sophistication. Several species practise group-spawning and communal defence of the resulting fry. Sometimes the fry from a previous brood remain within the parental territory to help protect younger siblings. At one time it was thought that only 'higher' vertebrates, such as birds or mammals, exhibited such 'advanced' behaviour.

All lamprologines are skilful jumpers and require tightly covered aquaria.

Above: **Julidochromis marlieri** ♂ *Striking body pattern.*

Julidochromis marlieri

● **Habitat:** A widely distributed, albeit uncommon, inhabitant of the rocky shore to depths of 35m (115ft).

● **Length:** Males up to 15cm (6in); females slightly less.

● **Sexing:** Males grow slightly larger than females and sport a conspicuous, penis-like genital papilla that inexperienced aquarists often mistake for an ovipositor.

● **Diet:** Micropredator.

● **Compatibility:** One pair defends a territory about 60cm (24in) square. It is possible to keep several pairs in the same aquarium as long as each can secure a territory. Individuals may behave aggressively towards the closely related *J. regani*, but otherwise prove good neighbours to Tanganyikan cichlids of their own size or slightly smaller. *Julidochromis marlieri* typically ignores midwater-swimming tankmates.

● **Breeding:** These monogamous cichlids pair readily, although the pair bond is not as strong as in most other lamprologines. Keep a few schooling fish with isolated couples to act as target fish and reinforce the pair bond. Like other *Julidochromis*, this species can either produce a large clutch of up to 300 olive-green eggs every 4-6 weeks, or spawn 12-20 eggs every 7 to 10 days for a period of several months. If a pair is pulsing its reproductive output in this way, the breeding tank will soon contain young of several different sizes, since older fry are tolerated in the presence of younger siblings. Feed the fry on microworms for their first few days of mobile life and then on brineshrimp nauplii. Fry survival is enhanced in an established aquarium containing a mature sponge filter. The young are sexually mature 14 months after spawning, at 7.5cm (3in) for males, slightly less for females.

The second large representative of the genus, *J. regani*, has the lateral stripes of *J. marlieri* but lacks the vertical bars. The maintenance requirements and reproductive patterns of the two species are identical. The closely related *Chalinochromis* species respond to the same care.

Pairs of the three dwarves of the genus, *J. dickfeldi*, *J. ornatus* and *J. transcriptus*, as well as dwarf species such as *Telmatochromis bifrenatus* and *T. vittatus*, will thrive in tanks 60cm (24in) long. All these species are about 7.5cm (3in) long and require the same care as *J. marlieri* in captivity.

Lamprologus brichardi

● **Habitat:** Large schools of this midwater-swimming species live in close association with rock faces that drop off steeply into the lake's depths. Though individuals can be found as deep as 30m (98ft), most reproductive activity takes place much nearer to the surface.

● **Length:** Males up to 9cm (3.5in); females slightly smaller.

● **Sexing:** Females are smaller than males and have a less rounded cranial profile. Length of fin filaments is not a reliable sexual distinction.

● **Diet:** Micropredator.

● **Compatibility:** Aggressive towards own species other than members of its extended 'family' in captivity, as well as towards other *Lamprologus* of a similar appearance. Otherwise, sexually inactive individuals make good community residents with companions too large to be easily swallowed. Parental fish defend a territory about 45cm (18in) square and are quite capable of killing tankmates unwilling, or unable, to respect its limits. The easiest way to secure a pair is to raise a group of young to maturity together. It is not unusual for this normally monogamous species to shift to harem polygyny if a surplus of females is present. Communal spawning – which occurs frequently in nature – is also likely if siblings are reared to maturity in a large, well-stocked community setting, where a single pair might encounter difficulties establishing a breeding territory. Sexual bonds tend to be stable even in the absence of target fish. Spawns can number 200 ovoid, olive-green eggs, but clutches of 50-75 are more normal. Fry are permitted to remain within their parents' territory, where they assist in the defence of their younger siblings, until they are sexually mature, eight months after spawning, at about 4cm (1.6in).

An albino form of *L. brichardi* is commercially available and several distinctive *L. brichardi*-like fish have been exported from different regions of the lake. Whether the 'black-faced', 'daffodil' and 'Kasagere' *Lamprologus* represent geographic races or subspecies of *L. brichardi*, or distinct species, remains to be determined. Take care to prevent hybridization between these attractive cichlids. *L. pulcher* is a superficially similar species with more pronounced rusty orange lateral spots and fin markings. More aggressive than *L. brichardi*, but shares the features of its reproductive behaviour.

Below: **Lamprologus brichardi** *Elegant flowing finnage.*

Lamprologus calvus

Pearly compressiceps

● **Habitat:** Found in close association with rocky bottoms to a depth of 12m (40ft). All known collecting sites are along the lake's southern and eastern shores.
● **Length:** Males up to 13cm (5in); females up to 9cm (3.5in).
● **Sexing:** Males grow larger and have a more massive head, as well as deeper dorsal and anal fins.
● **Diet:** Micropredator.
● **Compatibility:** Will eat fish as large as a female guppy, but poses no threat to tankmates too large to swallow. However, it is intolerant of others of its own species and of the superficially similar *L. compressiceps,* but usually gets along well with other retiring Tanganyikan cichlids. A slow-moving fish, easily intimidated by more assertive tankmates in a community tank, *Lamprologus calvus* is often the loser at feeding time. Digs less than other representatives of this group.

Like *L. leleupi* (page 123), males of this species will practise harem polygyny in captivity when the opportunity arises. Females prefer to deposit their eggs on the vertical wall of a cleft in the rockwork, but will accept a flowerpot or similar enclosed space as a spawning site in captivity. Spawns can number up to 100 whitish eggs, but half that number is closer to the norm. A significant percentage of the eggs often fails to develop. Whether this is the result of male failure to fertilize the entire clutch, or the female's rather casual hygienic behaviour towards the zygotes is unclear. Growth is slow, but *L. calvus* begins breeding well before it attains its maximum size. The young are sexually mature at about 12 months after spawning, at 5.7cm (2.25in) for males, 3.75cm (1.5in) for females.

The closely related *L. compressiceps* is somewhat deeper bodied than *L. calvus* and lacks its profusion of white dots. However, the maintenance requirements and reproductive pattern of these two species do not differ significantly.

Below: **Lamprologus calvus** *An inept predator on smaller fish.*

Above: **Lamprologus leleupi** *Brilliantly coloured, medium-sized cichlid.*

Lamprologus leleupi

● **Habitat:** Found in close association with the bottom over rocky shores to a depth of 70m (230ft). Relatively uncommon in water less than 40m (130ft) deep.
● **Length:** Males up to 10cm (4in); females up to 8cm (3.2in).
● **Sexing:** Females are smaller than males, with slightly shorter vertical fins and ventrals.
● **Diet:** Micropredator. Offer colour food regularly.
● **Compatibility:** Extremely aggressive towards its own species. Do not keep more than a single pair in tanks less than 150cm (60in) long. Will prey on tankmates the size of a male guppy. Sexually inactive individuals are otherwise good neighbours in a Tanganyikan community tank. Parental fish behave in the same manner as *L. brichardi* (see page 121).

Pairs breed readily if handled in the manner recommended for monogamous cichlids. In a large tank, a single male will usually spawn with all the available females, but generally involves himself in caring only for the last brood of fry he has sired. Spawns can number up to 200 greenish white, ovoid eggs. Unlike *L. brichardi*, adult fish do not tolerate the presence of older fry in their territory with the onset of another bout of reproductive activity, so be sure to separate parents and offspring 6-8 weeks after spawning. The young grow more slowly and less evenly than do *L. brichardi* fry, so sort them by size to prevent sibling cannibalism. Sexual maturity is attained 18 months after spawning, at a length of 6.5cm (2.5in) for males, slightly less for females.

A subspecies, *Lamprologus l. longior*, is often commercially available. It is characterized by its intense, constant yellow-orange coloration and somewhat slenderer build. *L. l. leleupi* is a somewhat stockier, golden ochre fish that turns intense lemon yellow when sexually active.

Above: **Lamprologus sexfasciatus** *A strikingly marked species.*

Lamprologus sexfasciatus

● **Habitat:** Found in close association with rocky bottoms to depths of 5m (16ft).
● **Length:** Males up to 14cm (5.5in); females identified as slightly smaller.
● **Sexing:** Females are somewhat smaller and fuller-bodied than males, but the only reliable means of sexing this species is by direct examination of the genital papillae.
● **Diet:** Micropredator.
● **Compatibility:** As described for *L. leleupi*, but given its larger adult size, *L. sexfasciatus* requires roomier quarters. A single pair will live happily in an aquarium 90-120cm (36-48in) long.

Handle this species in the manner recommended for *L. leleupi* (see page 123). It is not unusual for a pair of this slow-maturing species to go through several false spawns before actually producing a clutch of several hundred ovoid, off-white eggs. Although they appear infertile, the eggs are perfectly viable. Offer the very small fry microworms for the first few days, then *Artemia* nauplii. Fry survival seems greater if a mature sponge filter is present in the breeding tank for the fry to browse over. Remove the young from the breeding tank by the sixth week after spawning. They reach sexual maturity after about 18 months, at 9cm (3.5in) for males, 7.5cm (3in) for females.

Lamprologus signatus

● **Habitat:** Lives in close association with empty shells of the snail *Neothauma*, which occur in extensive drifts on open, sandy bottoms at depths of 10-100m (33-330ft).
● **Length:** Males up to 5cm (2in); females up to 3cm (1.2in).
● **Sexing:** Males of this species grow much larger than females and have deeper and more strikingly marked vertical fins.
● **Diet:** Micropredator.
● **Compatibility:** Males defend a territory about 30cm (12in) square against others of their sex. Several males can coexist as long as the tank is large enough to afford each a territory. As with most harem polygynists, females defend a much smaller area, restricted to

the vicinity of their shell. This species tolerates other Tanganyikan cichlids of the same size, or slightly larger, outside periods of sexual activity. However, parental fish become extremely intolerant of other bottom-dwelling fish. Easily bullied by larger lamprologines. Does best when kept exclusively with midwater-swimming tankmates too large to be swallowed.

This species will thrive equally well whether maintained in pairs or harems. Sexual bonding occurs freely, any mature individual spawning with another of the opposite sex but, unlike most lamprologines, female shell-dwellers display actively to the male before spawning. *L. signatus* insists on an empty snail shell as a spawning site. Eggs are placed deep within the shell, well out of sight, so female reluctance to leave its interior following a bout of courtship is the only indication that spawning has occurred. Up to 100 tiny fry emerge from the shell a week later and are allowed to forage near the opening under their mother's supervision. Offer them infusorians and microworms during the first week. Fry survival is always greater in a well-established aquarium containing a mature sponge filter. The female herds the fry back into the shell at night and the male blocks its aperture with sand and mounts guard over his immured family until morning, when he removes the plug and moves off to provide perimeter defence. The female continues to protect the fry for up to three weeks. She then chases them from her territory and eats any that remain behind. This can provoke an attack by the male, whose protective tendencies persist longer than the female's. To avoid intersexual conflict, separate the parents from their offspring at this point. Even at this early age, males are 50 per cent larger than females. The young reach sexual maturity 10 months after spawning, at 4cm (1.6in) for males, 2cm (0.8in) for females.

Below: **Lamprologus signatus** ♂ *An appealing little shell-dweller.*

Tanganyikan mouthbrooding cichlids

Although the species in this category stem from several different evolutionary lineages, they are for the most part maternal mouthbrooders (ie they incubate their eggs in the throat sac) with an openly polygamous mating system and should be treated accordingly (see page 76 of the section dealing with *Breeding and rearing*). The goby cichlids (*Eretmodus*, *Spathodus* and *Tanganicodus*) are monogamous mouthbrooders and both sexes share the task of incubation. They require the same care as any other monogamous cichlid (see page 80 for advice on breeding these cichlids).

Cyathopharynx furcifer

● **Habitat:** Schools of this social species are found at depths of 9-12m (30-40ft) just off rocky slopes.
● **Length:** Males up to 20cm (8in); females up to 16.5cm (6.5in).
● **Sexing:** Territorial males are instantly recognizable by their scintillating blue coloration. Subordinate males differ from females in their duskier overall coloration and longer ventral fins.
● **Diet:** Micropredator.
● **Compatibility:** Despite its rather large size, this species is quite unaggressive towards both its own species and other fish. It is possible to keep several males in a tank at least 180cm (72in) long. However, even in an aquarium of this size only one male will ever display its magnificent courting dress. Easily bullied by smaller, more assertive cichlids, this species is best housed alone or in the company of midwater-swimming, non-cichlid companions. Extensive pit digging by the dominant male makes it difficult to keep any but potted plants in its tank.

In nature, males build an elaborate crater nest by carrying sand to the top of rocky outcrops. In captivity, they simply excavate the largest pit possible at one end of the tank.

Below: **Ophthalmochromis nasutus** *One of the delicate featherfins.*

Courtship is prolonged; the male darts out from his nest, displays to the female and darts back again. Eventually she follows him to the pit and deposits a few eggs, which she immediately picks up. She then mouths the light tips of the male's ventral fins, which are folded back to the genital papilla during spawning. Clutches of up to 40 eggs have been reported, but two dozen is more usual. At 25°C (77°F), the incubation period is 21 days. Females often prove unreliable mothers and it may be helpful to offer them more privacy; if this does not have the desired effect, the only option may be to incubate the zygotes artificially. The newly released fry can take brineshrimp nauplii for their initial meal. They share the sensitivity of *Xenotilapia ochrogenys* to nitrogen cycle mismanagement and need similar care. The young begin breeding 10-12 months after spawning, at a length of 10cm (4in) for males, 9cm (3.5in) for females. However, up to 95 per cent of the eggs fertilized by such young males prove infertile. Males do not attain full reproductive competence until they are at least 18 months old.

Four different geographic colour forms of *C. furcifer* are known, and fishkeepers should prevent hybridization between them. All the Tanganyikan featherfins, which consist of the closely related genera *Aulonacranus*, *Cardiopharynx*, *Cyathopharynx* and *Ophthalmotilapia*, require the same care in captivity. These delicate species are best attempted by more experienced cichlid keepers.

Cyphotilapia frontosa

● **Habitat:** This species is restricted to rocky habitats. Juveniles are found at depths between 18 and 50m (60-164ft), but adults are rarely encountered in water less than 27m (90ft) deep.
● **Length:** Males up to 35cm (14in); females up to 25cm (10in).
● **Sexing:** Males are larger, with long, flowing fins and a very well developed nuchal hump.
● **Diet:** Piscivore.
● **Compatibility:** Males are extremely intolerant of one another. Do not house more than a

Below: **Cyphotilapia frontosa** *Keep in a spacious aquarium.*

single male per tank, even in aquaria more than 180cm (72in) long. A male is also apt to injure females if the fish are maintained in crowded quarters. A 150cm (60in)-long tank is the absolute minimum for a trio of *C. frontosa*. Tankmates that are too large to be easily swallowed are ignored. Despite its large adult size, this slow-moving species is easily intimidated by smaller, more aggressive cichlids, such as mid-sized *Lamprologus*. It does best when kept with other Tanganyikan mouthbrooders of a similar temperament. As a species it neither digs nor molests rooted plants.

Well-conditioned fish spawn readily if managed like any other openly polygamous cichlid. After a rather desultory courtship, the female, followed closely by the male, deposits a row of eggs on a solid surface. Only after the male has fertilized them does the female pick up the eggs. Well-fed females can produce as many as 60 large, off-white eggs in a spawning, but 20-25 is closer to the norm. The incubation period is 28 days at 27°C (80°F). This species does not care for its fry once they are released. Young females have a reputation for maternal unreliability and even some experienced females consistently devour their clutches. If isolating the egg-laying female fails to correct this problem, the only alternative is to use an 'artificial mouth' to bring the zygotes to term. The robust young can manage brineshrimp nauplii, sifted *Daphnia* and powdered prepared foods immediately after emerging from the female's mouth. With good feeding and frequent small water changes, they grow rapidly and reach sexual maturity 10-12 months after release, at 15cm (6in) for males, compared to 10cm (4in) for females.

Cyprichromis nigripinnis

● **Habitat:** A pelagic species found offshore on rocky slopes over a depth range of 10-50m (33-164ft).
● **Length:** Males up to 10cm (4in); females up to 9cm (3.5in).
● **Sexing:** Females are smaller than males and lack their dusky, iridescent blue-edged vertical fins.
● **Diet:** Micropredator.
● **Compatibility:** Always keep this extremely gregarious species in groups of six or more in captivity. Male aggression is highly ritualized, so it is usually possible to maintain several males together if they are not crowded. The fish require plenty of swimming space; a 150cm (60in)-long tank is the minimum recommended for any *Cyprichromis* species. Easily bullied by more assertive companions, but the smaller *Julidochromis* species and the several genera of goby cichlids make good tankmates, as do *Callochromis* and *Xenotilapia*. Neither digs nor molests rooted plants. Keep this accomplished jumper in a tightly covered tank.

Below: **Cyprichromis nigripinnis** ♂ *A pelagic 'sardine cichlid'.*

Above: **Tanganicodus irsacae** *Smallest Tanganyikan goby cichlid.*

Males actively court ripe females and contend vigorously for their attention. This is the only time that their aggressive behaviour may lead to injuries. After reciprocal circling, the female expels an egg and immediately dives down to seize it. The male then tilts slightly to one side and allows the female to mouth his vent. Spawns rarely exceed 20 eggs and are usually smaller. Egg-carrying females do not always display a bulging throat, but can be identified from above by their slightly outspread gill covers. In a single-species breeding tank, it is not necessary to isolate the female, since adults ignore newly released fry. The incubation period lasts 21 days at 25°C (77°F). The fry are large enough to take brineshrimp nauplii immediately on release. Because of their short intestinal tract and extremely rapid metabolism, they require numerous small daily feeds to prosper. They are also extremely sensitive to dissolved metabolites. The young attain sexual maturity 10-12 months after spawning, at 7cm (2.75in) for males, 5.7cm (2.25in) for females. *C. microlepidotus* and *C. leptasoma* require the same care in captivity.

Tanganicodus irsacae

● **Habitat:** This bottom-living species is found in the surge zone of cobble beaches. Like other Tanganyikan surf-dwellers, it lacks a functional swimbladder and moves over the bottom with droll hopping motions.

● **Length:** Males up to 9cm (3.5in); females slightly less.

● **Sexing:** Apart from the marginal difference in size, there is no way of distinguishing between the sexes.

● **Diet:** Micropredator. Swims with difficulty; provide foods that sink to bottom.

● **Compatibility:** Pairs are very aggressive towards companions of their own species and have little use for the close company of other goby cichlids. Despite its small adult size, it is not possible to keep more than a single pair in a tank 120cm (48in) long. This species is easily bullied by larger *Lamprologus* species, but otherwise gets along well with a wide range of Tanganyikan cichlid companions. Like other goby cichlids, it poses no threat even to much smaller midwater-swimming tankmates. Not a digger, but a skilful jumper; provide a tightly covered tank.

The best way to secure a pair is to rear a group of young to maturity together and, as soon as two fish begin to act like a pair, remove the remaining specimens. Courtship is desultory and followed by reciprocal circling, during which the eggs are deposited and fertilized. One of the spawning fish picks up the flattened, discoid eggs and carries them for half of the incubation period. Then its partner takes over the task of brooding the zygotes. Spawns range from 6 to 20 eggs. Because so few eggs do not produce a conspicuous distension of the throat and the egg-laying parent continues to feed during the incubation period, it is easy to miss a spawning. Often the first indication that breeding has occurred is the appearance of fry among the tank's rockwork. At 27°C (80°F), the young are released 21 days after spawning. There is no parental care once the fry have been released; in fact their parents ignore them. The young can take brineshrimp nauplii and powdered prepared foods for their initial meal. They are extremely sensitive to any build-up of metabolic waste.

However, if kept under a regime of frequent small partial-water changes, they are not difficult to rear. It takes the young a year to attain full adult size, and an additional 12 months to attain sexual maturity.

Tropheus moorii

● **Habitat:** Found over rocks supporting dense algal growth. Adults are seldom encountered at depths greater than 10m (33ft).
● **Length:** Males up to 14cm (5.5in); females up to 10cm (4in).
● **Sexing:** Males are larger than females and sport a distinctly 'Roman' nose, but direct examination of the genital papillae is the only reliable means of sexing this species.
● **Diet:** Herbivore.

Below: **Tropheus moorii** *Available in many colour forms.*

● **Compatibility:** Males of this particular species are extremely intolerant of one another. Avoid introducing mbuna tankmates, but smaller lamprologines and midwater-swimming companions are tolerated. To thrive in captivity, a group of *T. moorii* require a tank at least 120cm (48in) long. This species will eat soft-leaved aquatic plants, but does not dig.

● **Breeding:** Like other openly polygamous cichlids, this species spawns readily when maintained in single male/multiple female groups. The male does not display to a ripe female from a fixed territory, but follows her about the tank. The pair drop to the bottom and start to circle one another. The female expels the large eggs one at a time and immediately picks them up. They are fertilized in her mouth as she nips directly at the male's vent. Broods of up to 20 fry have been reported in captivity, but a dozen eggs per spawning is closer to the norm. The incubation period is 28 days at 27°C (80°F). Well-developed brood care is characteristic of all *Tropheus* species and females exercise caution before releasing their progeny for the first time. This may account for observations that the incubation period lasts 30 days in *Tropheus* species. If the breeding group has in its own tank, there is no need to separate the parental female from her companions, since adequately fed adults ignore newly released fry. The robust fry can immediately take brineshrimp nauplii and powdered foods and quickly emulate the grazing behaviour of their elders. With due attention to good aquarium hygiene, they are easy to rear and grow rapidly. The young are sexually mature 10-12 months after spawning, at 7.5cm (3in) for males, slightly less for females.

There are over two dozen recognized geographical colour forms of this species. Some of these refuse to interbreed in captivity and may actually constitute distinct biological species. As a general rule, you should make every effort to avoid crossing these colour forms.

Xenotilapia ochrogenys

● **Habitat:** Lives in large schools over open, sandy substrates at depths not exceeding 20m (66ft). Breeding groups usually inhabit much shallower water, no more than 4m (13ft) deep.

● **Length:** Males up to 15cm (6in); females 10cm (4in).

● **Sexing:** Males are easy to distinguish by their dazzling colours and breeding dress.

● **Diet:** Micropredator.

● **Compatibility:** Males defend a territory about 60cm (24in) square against other males, but several males will coexist if the tank is large enough to afford each a territory. Although easily bullied by more aggressive cichlids, *X. ochrogenys* does get along well with *Cyprichromis* species or any other midwater-swimming fish. This species does not eat aquatic plants, but its habit of moving

Above: **Xenotilapia ochrogenys** ♂ ♀ *A courting pair (male below).*

massive amounts of gravel when digging spawning pits places all save potted specimens in continual jeopardy.

Xenotilapia ochrogenys spawns freely if kept in multiple female groups in a tank of its own. Spawning is preceded by vigorous courtship of the female and intense pit digging by the male. The female deposits and picks up a few eggs after a period of reciprocal circling and vent nudging. Clutches range from 5 to 30 eggs. It is easy to overlook a spawning; such small clutches do not produce a marked throat bulge and females will continue to feed, albeit at a reduced rate, while carrying them. Young females have a reputation for maternal unreliability, and even older individuals may abort the incubation sequence if the breeding tank is too small for them to avoid the male's attentions. At 27°C (80°F) the zygotes are carried for 21 days. The newly released fry can take brineshrimp nauplii and fine powdered food for their initial meal. They are extremely sensitive to accidental pollution of their tank and punctilious attention to proper nitrogen cycle management is vital. In order to avoid damaging fights between maturing males, it is a good idea to segregate the sexes as the young begin to manifest sexual colour differences. This is possible at 6-8 months after release, but the young fish do not attain reproductive competence until they are a year old, at 7.5cm (3in) for males, and slightly smaller for females. It takes them an additional year to grow to their full adult size.

This dazzling species is representative of the many substrate-sifting cichlids of the genera *Callochromis* and *Xenotilapia*. Their scintillating colours and reasonably mellow dispositions make them an obvious choice for the fishkeeper, but until recently they have enjoyed limited availability. Adults are very difficult to ship successfully because of their extreme sensitivity to dissolved metabolites and their high dissolved oxygen requirements. Juveniles are much better travellers, so with the increasing availability of tank-reared fry, these species can be expected to gain the popularity amongst enthusiasts that they deserve.

LAKE MALAWI CICHLIDS

Like their Tanganyikan counterparts, these fish are all acutely intolerant of dissolved metabolites. Efficient biological filtration and frequent partial water changes are absolutely essential to their well-being in captivity. To minimize aggression under aquarium conditions, maintain these cichlids at relatively high population densities, but high stocking rates are practical only when scrupulous attention is paid to nitrogen cycle management. Malawian cichlids require hard, alkaline water to prosper. A temperature range of 22-25°C (72-77°F) will suffice for day-to-day maintenance, with an increase to 27-30°C (80-86°F) for breeding. These cichlids are capable jumpers, so keep them in a tightly covered tank. Most mbuna eat soft-leaved plants, while the males of the majority of Malawian cichlids move enormous volumes of gravel when constructing a nest pit.

These fish are all openly polygamous maternal mouthbrooders. Managed correctly, they breed readily in captivity. Spawning follows a prolonged and often vigorous courtship. After a period of reciprocal circling, the fish spawn in the typical haplochromine manner. Although the eggs are fertilized within the female's mouth, she does not elicit ejaculation by nipping at the pseudo-ocelli often present on the male's anal fin. Her target is the male's vent, where his genital papilla often contrasts strongly with the background colour of the abdominal region. At 30°C (86°F), the incubation period lasts 21 days. The robust fry can take brineshrimp nauplii and powdered food for their initial meal and are easy to rear. When selecting future breeding stock remember that all male haplochromines grow more quickly than females.

Sexually active male Malawian cichlids are strongly territorial. In captivity, favourable living conditions promote year-round male sexual activity, while space limitations preclude the natural expression of typical territorial behaviour. This is why, with rare exceptions, it is not possible to house several males of the same species together in the presence of females.

The unavoidable crowding of these fish under aquarium conditions also appears to explain the phenomenon of hyperdominance, in which one male will harass males of other species so severely that he prevents them from expressing any sort of territorial or courtship behaviour. In the absence of appropriate sexual cues from their own males, females of other species eventually respond positively to the hyperdominant male's courtship, producing unwanted but often viable hybrids.

Ideally, Malawian cichlids are best kept in 'harems' of a single male and several females, one species to a tank. In practice, however, most aquarists prefer to maintain a mixed-species community of cichlids. To reduce the risk of unwanted hybridization, observe the following guidelines:

- Do not house together species with markedly different aggressive tendencies. The male of the most aggressive species will become hyperdominant.

- Do not house together very closely related species, or those with very similar male courting dress. Females are less likely to succumb to the overtures of a very differently coloured male of another species.

● Do not house together species whose females are very similar Fry initially resemble their mother, not their father. The results of hybridization between two species with identical female colours will not be recognized until males begin to colour up. However, if the females of the hybridizing species have very different colour patterns, hybrid fry are easy to recognize.

Hyperdominance is chiefly a consequence of limited living space. The larger the tank available for a community of Malawian cichlids, the less the likelihood of unwanted hybridization. If these guidelines are followed, most of the popular Malawian species in the 7.5-15cm (3-6in) range can be housed in tanks 150-180cm (60-72in) long.

Mbuna compatibility

Group 1

Species that will behave aggressively towards other fish in tanks more than 180cm (72in) long and are likely to kill or injure tankmates in smaller aquaria.

Genyochromis mento
Melanochromis chipoke
M. loriae
M. melanopterus
M. simulans
M. 'lepidophage'
All *Petrotilapia* spp.
Pseudotropheus brevis
Ps. crabro complex
Ps. elongatus complex
Ps. hajomaylandi
Ps. lucerna
Ps. tursiops
Ps. williamsi complex

Group 2

Species that usually forgo serious aggression in tanks more than 180cm (72in) long; large specimens are likely to injure or kill other fish in tanks less than 150cm (60in) long.

Labeotropheus fuelleborni
Melanochromis auratus
M. parallelus
M. vermivorous
Pseudotropheus aurora
Ps. greshakei
Ps. heteropictus
Ps. cf. *livingstonii*
Ps. lombardoi
Ps. zebra complex

Group 3

Species that usually forgo serious aggression in tanks more than 150cm (60in) long; large specimens can pose some risk to their companions in tanks less than 120cm (48in) long.

Cyathochromis obliquidens
All *Gephyrochromis* spp.
Labeotropheus trewavasae
Melanochromis interruptus
M. johanni complex
Pseudotropheus barlowi
Ps. elegans
Ps. gracilior
Ps. lanisticola
Ps. macrophthalmus complex
Ps. cf. *microstoma*
Ps. minutus
Ps. socolofi complex
Ps. tropheops complex

Group 4

Species that usually forgo serious aggression in tanks more than 120cm (48in) long.

All *Cynotilapia* spp.
Iodotropheus sprengerae
All *Labidochromis* spp.

Mbuna

The Tonga people, who live on the western coast of Lake Malawi, apply the term 'mbuna' (rockfish) to representatives of a distinctive group of haplochromines defined by a number of morphological and behavioural characteristics. It is easy to see why the mbuna were the first cichlids exported as aquarium fish. Most species range from 6.5-13cm (2.5-5in), have dazzling male and, in some instances, female coloration, are vivacious in manner and easy to maintain. They have been compared to marine fish, such as the damselfish and wrasses. The comparison is apt, for mbuna are both brightly coloured and aggressive. To keep them successfully it is important to understand their aggressive behaviour.

As with most other cichlids, mbuna aggression is related to the defence of a territory and this territoriality is usually a function of reproduction. Mbuna differ in that males of many species defend a multipurpose territory that provides food, shelter from predators and a spawning site. In a few instances, even females defend a feeding territory against potential competitors of the same and other species. Obviously, the value of such territories persists well beyond the duration of a single spawning episode, hence the phenomenon of persistent territorial defence in nature.

Below: **Pseudotropheus zebra**
The striking coloration of the male has assured its popularity.

The size of the area defended by a resident fish in nature is so large that in captivity it is seldom practical to think of housing more than a single male of a given species per tank, even in well-furnished aquaria as large as 180cm (72in) long. Matters are further complicated by the fact that a territory holder will attack a similarly coloured species, or one with a very similar feeding pattern, just as vigorously as an intruder of the same species. However, it is possible to avoid serious aggression in a mbuna tank by carefully selecting tankmates that do not perceive one another as threatening.

The table divides the most widely available mbuna species into four groups in decreasing order of their aggressiveness under aquarium conditions. While making allowances for the individuality of cichlids in general, it should still be possible to assemble a reasonably harmonious community of these cichlids, using these guidelines.

- Representatives of adjacent groups, eg groups 2 and 3, are more likely to coexist separated than those in disconnected groups, eg 1 and 3.

- Among cichlids in the same group, fish of the same genus are more likely to prove incompatible than representatives of different genera.

- Among representatives of the same genus, species with similar body shapes are more likely to prove incompatible than those that differ in this regard.

- Regardless of the group to which they belong, species with similar male courting coloration are apt to prove incompatible when housed in the same aquarium.

Mixing mbuna with other Malawian haplochromines is not desirable, nor should you house mbuna with their Tanganyikan cousins – *Petrochromis*, *Simochromis* and *Tropheus* species – since the latter will not prosper in their company. Well-fed mbuna typically ignore midwater-living companions too large to make a convenient mouthful, but their own vivacity usually renders dither fish superfluous in their tank.

Mbuna breed freely in a community setting. With the exception of the long-snouted *Melanochromis* species, which prey efficiently on newly released cichlid fry, adults are disinclined to molest either their own young or those of other species. Assuming their aquarium is well furnished with rockwork and free of potential predators, mbuna fry have a good chance of growing to maturity in a community of their elders, as long as they are fed. All mbuna thrive on the diet recommended for herbivorous fish. Do not offer them *Tubifex* worms in any form or frozen brineshrimp. The coloration of many species responds to colour foods.

Mbuna species were originally defined on the basis of morphological characteristics. However, each isolated stretch of shoreline or offshore island group supports its own distinctive colour phenotypes and ichthyologists disagree over whether these should be treated as regional colour variants of a single species, such as *Pseudotropheus tropheops*, or as valid, albeit closely related, species. Until more is known about the genetics and behaviour of these often very distinctive phenotypes, aquarists are well advised to treat them as distinct species and to avoid crossing them.

Iodotropheus sprengerae

Rusty cichlid

● **Habitat:** Found over a wide range of rocky habitats and sometimes even in the intermediate zone at Boadzulu, Chinyamwezi and Chinyamkwazi Islands. Although individuals have been recorded in water 40m (131ft) deep, this species is most abundant at depths of 3-15m (10-50ft).

● **Length:** Males up to 9cm (3.5in); females up to 7cm (2.75in).

● **Sexing:** Both sexes are the same colour and can be distinguished only by the presence of large, clearly defined pseudo-ocelli on the male's somewhat longer anal fin.

● **Diet:** Micropredator.

● **Compatibility:** Males are territorial for a brief period before spawning, and then only towards one another. It is thus possible to keep several males in a tank more than 120cm (48in) long. Do not house *I. sprengerae* with more belligerent mbuna.

Below: **Iodotropheus sprengerae** ♂ *Hardy and unaggressive.*

Above: **Labeotropheus fuelleborni** *One of many colour forms.*

● **Breeding:** Spawns range from 5 to 60 eggs. This species does not undergo a colour metamorphosis and displays remarkable reproductive precocity. The young can begin breeding as early as the 14th week after spawning, at 3.2cm (1.25in) for males, 2.5cm (1in) for females. They do not attain full adult size until they are at least 8-10 months.

The only species with which the rusty cichlid is likely to be confused is *Labidochromis vellicans.* Both species share a rusty base colour, but *L. vellicans* has a more pointed snout. By virtue of its relatively mellow temperament, attractive coloration and willingness to breed in captivity, *Iodotropheus sprengerae* is the ideal mbuna for a beginner.

Labeotropheus fuelleborni

● **Habitat:** Abundant over areas of broken rock and in the intermediate zone throughout the lake. Most abundant in water less than 5m (16.5ft) deep. Individuals can penetrate 18m (60ft).
● **Length:** Males up to 12cm (4.7in); females slightly smaller. Captive specimens often reach 15cm (6in).
● **Sexing:** Males grow larger, and their colour metamorphosis begins at 6.5cm (2.5in).
● **Diet:** Herbivore. Appreciates fresh vegetables.
● **Compatibility:** Males are extremely aggressive towards males of their own species but tend to disregard males of other species. A significant exception to this rule are males of *L. trewavasae.* Indeed, efforts to house these very similar species together in captivity usually end badly for one or the other. Very large males are apt to become hyperdominant in tanks 180cm (72in) long.
● **Breeding:** Males are likely to harass egg-carrying females, whose spawns can number 75 eggs. The onset of sexual activity occurs 9-12 months after release, at 7.5cm (3in) for males, slightly less for females.

Like *Ps. zebra*, this species is characterized by both colour polymorphism and extensive geographic colour variation. Over 20 distinct races of *L. fuelleborni* have been reported in nature and several of these have been exported. Fortunately for aquarists, the available evidence argues against the existence of cryptic biospecies (ie biologically distinct fish that do not differ in shape, but often differ in details of coloration) of the sort that make up the *Ps. zebra* complex. True-breeding strains of the orange/black morph are also available.

Above: **Labeotropheus trewavasae** *Offer fresh vegetable foods.*

Labeotropheus trewavasae

● **Habitat:** Most common over large rocks, but has been recorded from other habitats throughout the lake. Individuals are fairly evenly distributed to depths of 20m (66ft).
● **Length:** Males up to 12cm (4.7in); females slightly smaller. Captive specimens often reach 15cm (6in).
● **Sexing:** Males begin to change colour at 6.5cm (2.5in).
● **Diet:** Herbivore.
● **Compatibility:** Males are less markedly territorial than *L. fuelleborni*, but do not house the two species together (see *L. fuelleborni* page 137). *L. trewavasae* seldom displays hyperdominance in captivity.

Sexual maturity is reached 9-12 months after release, at 7.5cm (3in) for males, slightly less for females. Spawns range from 10 to 60 eggs.

L. trewavasae is characterized by both colour polymorphism and extensive geographic colour variation. Several of the 13 recognized races have been exported. Of these, the most striking is the Chitende population, whose males are rusty ochre with icy mauve vertical fins. The Tumbi Island race is characterized by both orange/black and solid orange female morphs. A strain in which both sexes are orange and true-breeding orange/black strains are commercially available.

Labidochromis zebroides

Likoma Island clown

● **Habitat:** The interstices of large and medium-sized rocks at Likoma Island. This shallow-water species is always found at depths of less than 6m (20ft).
● **Length:** Males up to 6.5cm (2.5in); females up to 5cm (2in).
● **Sexing:** Females are a uniform silvery grey. Males begin to change colour at 2.5cm (1in).
● **Diet:** Micropredator.
● **Compatibility:** The behaviour of this species is typical of the genus as a whole. Males defend small territories, about 30cm (12in) square, based upon a cave or similar shelter. Given their small adult size and modest space requirements, it is possible to house several males in a tank more than 120cm (48in) long. Despite its small adult size, this species can be safely housed with larger mbuna, as long as the tank is well furnished and contains plenty of cover. This species does not manifest hyperdominant behaviour in captivity.

Sexual maturity coincides with the complete expression of male coloration, at about eight months after release, at 3.8cm (1.5in) for males, slightly less for females. Spawns range from 5 to 30 eggs. Unlike other mbuna species, female *Labidochromis* do not actually practise parental care of their fry.

Above: **Labidochromis zebroides** ♂ *A most beautiful mbuna.*

Labidochromis freibergi is a similarly coloured species from the Likoma Island group and is often available commercially. It has a shorter snout, more rounded cranial profile and deeper body. For no immediately obvious reason, *L. freibergi* is frequently sold under the name *Pseudotropheus minutus.*

Melanochromis johanni

● **Habitat:** Occurs over rocky habitats and in the intermediate zone along the eastern shoreline of the lake. This is a shallow water species; its distribution is restricted to water 6m (20ft) deep or less.

● **Length:** Males up to 9cm (3.5in); females slightly smaller. In captivity, males can grow up to 12.7cm (5in); females up to 10cm (4in).

● **Sexing:** In some strains, males begin to change colour at 2.5cm (1in).

● **Diet:** Omnivore.

● **Compatibility:** Young males tolerate each other well enough, but as they grow larger, it is virtually impossible to keep more than one male per tank. Best kept with other *Melanochromis* species, for its small adult size puts it at a severe disadvantage. This species seldom displays hyperdominance in captivity.

Below: **Melanochromis johanni** ♂ *A territorial male.*

Sexual maturity is attained at about eight months after release, at 6.5cm (2.5in) for males, 5cm (2in) for females. Spawns range from 7 to 60 eggs.

Two populations with identical juvenile coloration but different male and, in one instance, female colour patterns have been exported. Instead of the parallel stripes of *M. johanni*, the males of the Chisumulu Islands, are sooty black with two parallel rows of bright blue blotches on the flanks. The second population, restricted to Likoma Islands, is characterized by females marked with black and blue-white stripes. Until their status is resolved, it is best to treat both as distinct species.

Pseudotropheus lombardoi

● **Habitat:** Only found at Mbenji Island and most abundant at the rock/sand interface. Also occurs over large, flat rocks. Lives in water 2-25m (6.6-82ft) deep, but most abundant at 10m (33ft).
● **Length:** Males up to 10cm (4in); females slightly smaller. Like most mbuna, this species grows substantially larger in captivity.
● **Sexing:** Males begin to change their bright blue juvenile coloration to the yellow adult coloration at 4cm (1.6in).
● **Diet:** Herbivore.
● **Compatibility:** Males are less belligerent towards other mbuna than *Ps. zebra* (see page 143).

Male sexual maturity coincides with the completion of the colour metamorphosis, eight months after release, at 6cm (2.25cm). Females mature concurrently, at 5cm (2in).

Spawns range from 6 to 75 eggs, depending upon the size and condition of the female. These often lose their bright blue juvenile coloration after carrying several broods and become a pale beige with a blue wash on the flanks.

Pseudotropheus macrophthalmus 'yellow head'

● **Habitat:** Particularly common along rock/sand interfaces at Likoma and Chisumulu Islands. Specimens have been recorded as deep as 23m (75ft), but most abundant in the shallows, between 1-5m (3.2-16.5ft).
● **Length:** Males up to 11.5cm (4.5in); females grow up to 9cm (3.5in).
● **Sexing:** Females are silvery grey with a lilac wash and a row of small dark blotches along the midlateral line and the back. Distinctive male coloration begins to appear at 5cm (2in).

Below: **Pseudotropheus lombardoi** ♂ *Striking male coloration.*

Above: **Pseudotropheus macrophthalmus 'yellow head'** ♂.

● **Diet:** Herbivore.
● **Compatibility:** Males are very intolerant of their own species at all times and become progressively more aggressive towards other males with the approach of spawning. Not given to hyperdominant behaviour.

Reproductive pattern as outlined for the group. Brood sizes range from 7 to 60 fry. The young are sexually mature at about eight months after release, at 6.5cm (2.5in) for males, slightly less for females. The male's yellow coloration may not be fully developed for another 6 months.

This widely available representative of the *Ps. macrophthalmus* species complex displays considerable variability in coloration. Individual males in some local populations are almost entirely yellow. *Pseudotropheus macrophthalmus* 'red cheek' is a small-mouthed species that coexists with *Ps. macrophthalmus* 'yellow head' in the Likoma Island group. As its name implies, the colour in the head and shoulder region of adult males is a deeper orange, while females are a uniform golden yellow. A third, similarly coloured representative of this complex occurs along the southwestern shore of the lake. As its name implies, males of *Ps. macrophthalmus* 'orange chest' sport an intense yellow-orange flush in front of the belly and on the chest that extends to the opercula and cheeks. Females are greyish brown with seven or eight darker grey vertical bars on the flanks. An extremely attractive orange/black morph of this species is known. These representatives of the *Ps. macrophthalmus* species complex are also commercially available, as are the closely related *Ps. tropheops*, *Ps. gracilior* and *Ps.* cf. *microphthalmus*.

Pseudotropheus minutus

● **Habitat:** Over the upper surfaces of large boulders and rock slabs. Although its depth range extends to 40m (130ft), it is most abundant in water 3m (10ft) deep or less.
● **Length:** Males up to 10cm (4in); females up to 7.5cm (3in).
● **Sexing:** Young females retain the golden beige base colour of juveniles but, with age, they come to resemble duller versions of the male. Males begin to lose the golden yellow juvenile coloration at 3.8cm (1.5in) and attain their adult coloration at 5.7cm (2.25in).
● **Diet:** Herbivore.
● **Compatibility:** As for *P. socolofi* (see page 142).

Above: **Pseudotropheus minutus** ♂ *A territorial male.*

Reproductive pattern as outlined for the group. Females begin breeding 6-8 months after release, at 5cm (2in). A single spawning can yield from 5 to 50 fry.

This slender mbuna, sometimes known as the 'Likoma Island Elongatus', has largely displaced the true *Ps. elongatus* commercially. That species is one of the few mbuna in which both sexes ferociously defend a territory against allcomers, which doubtless accounts for its rather lurid reputation as an aquarium terrorist! Despite its brilliant coloration, *Ps. elongatus* is not a very satisfactory aquarium resident and its replacement by the equally colourful, but much mellower, *Ps. minutus* is hardly surprising.

Pseudotropheus socolofi

● **Habitat:** An uncommon resident of the rock/sand interface along the eastern coast of the lake at depths of 4-10m (13-33ft).
● **Length:** Males up to 10cm (4in); females slightly smaller.
● **Sexing:** Both sexes share the same powder-blue coloration, but males have larger, more clearly defined and, as a rule, more numerous yellow-orange pseudo-ocelli on their anal fins.
● **Diet:** Herbivore.
● **Compatibility:** Males are less aggressive towards other males of their own species than *Ps. lombardoi.* Males of this species defend a territory about 45cm (18in) square, which makes it possible to actually house more than a single individual in tanks 120-150cm (48-60in) long.
● **Breeding:** Reproductive pattern as outlined for the group. As many as 75 fry can be produced in a single spawning, but brood sizes of 35-50 are usually the rule. This species does not undergo a sexually related male colour change. The fry are exactly the same colour as their parents and require about eight months to reach sexual maturity, at 6.5cm (2.5in) for males, 5cm (2in) for females.

he adult colour pattern of *Ps. socolofi* is sufficiently similar to that of the undescribed species sold as *Pseudotropheus* 'kingsleyi' to account for confusion over the identity of these two mbuna. Males of *Ps.* 'kingsleyi' rarely exceed 8.3cm (3.25in). They sport an iridescent white or pale yellow spiny dorsal fin and dusky upper and lower distal margins in the caudal. Females are a uniform greyish beige. *Ps. lucerna* is another similarly coloured species, but larger (males up to 15cm/6in) and extremely aggressive.

Above: **Pseudotropheus socolofi** ♂ *Males have larger egg spots.*

Pseudotropheus zebra

● **Habitat:** Areas of broken rock at depths of 5-20m (16.5-66.5ft).
● **Length:** Males up to 12cm (4.75in); females slightly smaller. Captive individuals can reach 15cm (6in).
● **Sexing:** Males have clearly defined egg spots on the anal fin.
● **Diet:** Herbivore.
● **Compatibility:** Male *Ps. zebra* defend territories 2m (6.6ft) square in nature. They are very hard on similarly marked mbuna males and apt to become hyperdominant in tanks less than 180cm (72in) long.
● **Breeding:** Large males are given to post-spawning harassment of egg-carrying females, hence the importance of providing shelter. Spawns of 30-50 eggs per clutch are normal. Males begin their colour change at 4cm (1.6in). It is complete and sexual activity begins eight months after release, at 7.5cm (3in) for males 6.5cm (2.5in) for females.

Ps. zebra is characterized by both colour polymorphism and the existence of numerous geographic colour variants.

Below: **Pseudotropheus zebra** ♂ *Orange/black aquarium morph.*

Other Lake Malawi cichlids

Lake Malawi supports 177 described species of non-mbuna haplochromines, currently apportioned between a total of 38 genera.

By a conservative estimate, 79 of these representing 29 genera have been exported at least once over the past 30 years. Of this number, 64 species referable to 23 genera have, to my knowledge bred at least once in captivity.

Comprehensive coverage of these cichlids is clearly beyond the scope of this book; the aim has been to present a generally available, representative selection of *Haplochromis* and species of allied genera from several of the main ecological groupings. The great popularity of the 'Malawi peacock cichlids' warrants the inclusion of a wide, though by no means comprehensive, selection of *Aulonacara* phenotypes. All these cichlids require spacious aquarium tanks to flourish in captivity.

Aulonacara baenschi

Yellow peacock; sunshine peacock

● **Habitat:** The rock/sand interface in the Chipoka and Maleri Island groups and on the western coast of the lake at Nkhomo. Isolated males and small groups shelter under rock overhangs at depths of 8-20m (27-66ft).

● **Length:** Males up to 9.5cm (3.75in); females up to 7.5cm (3in). Captive males can reach 12.7cm (5in).

● **Sexing:** Males begin to change colour at 3.75cm (1.5in), but it may be another three months before males develop the full intensity of their yellow coloration.

● **Diet:** Micropredator. Offer colour foods regularly.

● **Compatibility:** As for *A. stuartgranti* (see page 146).

Sexual maturity is attained 6-7 months after release, at 5cm (2in) for males, slightly less for females. Spawns can number from 10 to 40 eggs. Yellow peacock fry are smaller and more delicate than those of *Aulonacara stuartgranti;* careful attention to nitrogen cycle management is essential

A longer-snouted *Aulonacara* with a metallic blue head and bright yellow body and fins is sold

Below: **Aulonacara baenschi** ♂ *Colour feeding is essential.*

Above: **Aulonacara freibergi** ♂ *Males are very aggressive.*

under the name yellow-sided peacock. A second, short-snouted species from Usisya, on the northwestern coast of the lake, is known to aquarists as the flavescent peacock. It, too, has a metallic blue head, but its flanks are golden orange rather than bright yellow, and both the dorsal and anal fins sport broad black submarginal bands. Although it may be a colour form of *A. baenschi*, it seems more likely that it represents another undescribed species. Both cichlids are somewhat more aggressive than the yellow peacock under aquarium conditions.

Aulonacara jacobfreibergi

Butterfly peacock

● **Habitat:** The interface between rocky and sandy habitats at numerous localities along the lake's extreme southern coast. Small groups or solitary territorial males shelter under rock overhangs at depths of 4-12m (13-40ft).

● **Length:** Males up to 10cm (4in); females up to 7.5cm (3in). Males grow as large as 15cm (6in) when kept in captivity.

● **Sexing:** Males begin their colour metamorphosis at 2.5cm (1in).

● **Diet:** Micropredator. Offer colour food regularly.

● **Compatibility:** As for *A. stuartgranti*, but this species is even more aggressive towards other males of the same species. Large specimens often behave in a hyperdominant manner, even in tanks over 150cm (60in) long.

This species resembles the regal peacock in its sexual precocity. Males can attain full adult coloration and sexual maturity as early as six months after release, at 3.8cm (1.5in). Females begin breeding at the same time, at 3.2cm (1.25in). Spawns of up to 55 eggs have been recorded, but broods of 25-30 fry are more likely.

A closely related, but slenderer peacock species, whose males sport a more deeply notched caudal fin, broader, iridescent white fin margins and well-developed anal fin pseudo-ocelli, has been exported from the southern part of the lake under the names *Aulonacara* 'caroli' and swallowtail peacock. Both these *Aulonacara* species are spectacular show fish.

Above: **Aulonacara stuartgranti** *Nkhomo Dwarf Regal Peacock.*

Aulonacara stuartgranti

Regal Peacock

● **Habitat:** The interface between rocky and sandy habitats at isolated localities along the western coast of the lake. Small groups shelter under rock overhangs at depths of 8-15m (27-50ft).

● **Length:** Males up to 9.5cm (3.75in); females up to 7.5cm (3in). Males can attain 14cm (5.5in) in captivity.

● **Sexing:** Full male coloration appears 6-7 months after release.

● **Diet:** Micropredator.

● **Compatibility:** Males are always intolerant of males of their own species and become progressively more aggressive towards other male haplochromines when spawning. They are also likely to become hyperdominant if housed with other *Aulonacara*. Utaka, the specialized zooplankton feeders, and the smaller chisawasawa that live close to sandy substrates, are the preferred companions for this or any other *Aulonacara* species.

Spawns can range from 10-50 eggs, but 20-30 is more normal. Males measuring 3.8-5cm (1.5-2in) at about six months and females of the same age can begin breeding. While such efforts are usually successful, these early broods rarely number more than 10 fry.

Above: **A. stuartgranti** ♀ *Female's prominent lateral bars are characteristic of the genus.*

The generic name of these cichlids, literally translated from the Greek, means 'flute-face'. It refers to the enlarged canals of the cephalic lateral line system, which resemble the openings of a flute. The enhanced sensitivity to movement afforded by the expanded lateral line system may allow these cichlids to locate their invertebrate prey more effectively in dim lighting. A second metallic blue *Aulonacara*, characterized by a bright orange-red girdle in the shoulder regions and orange ventral fins, is widely available. Originally sold as the red-shouldered peacock, this species has recently been described and given the name *Aulonacara hansbaenschi*.

Chilotilapia rhoadesi

● **Habitat:** Sandy and intermediate zones. Often associated with *Vallisneria* beds and abundant in water 3m (10ft) deep or less.
● **Length:** Males up to 25cm (10in); females up to 20cm (8in).
● **Sexing:** Males begin to change colour at 11cm (4.3in), 10 months after release. However, full male coloration is not expressed until the fish is two years old.
● **Diet:** Micropredator. Live snails are a treat.
● **Compatibility:** As outlined for *S. ahli*, although its requirements for living space are identical to those of *N. venustus* (see page 151). Though less aggressive than males of that species, there are reports of individual male *C. rhoadesi* becoming hyperdominant in a community tank. It is best not to house this fish with species whose females share the combination of an oblique subdorsal stripe and pronounced midlateral band, lest it prove impossible to identify any possible hybrid fry that might result from such cohabitation. Sexual maturity is attained at about 18 months, at 15cm (6in) for males, slightly less for females. Brood sizes range from 25 to 120 eggs, with clutches of 50-60 closer to the norm.

Despite its large adult size, *C. rhoadesi* is such a spectacular show fish that its continued popularity seems assured.

Sciaenochromis ahli

Electric blue haplochromis
● **Habitat:** A rare deep-water species found over rocks.
● **Length:** Males up to 18cm (7in); females up to 15cm (6in).
● **Sexing:** Males are slow to change colour. The process begins with the appearance of metallic blue in the facial region at 7.5cm (3in) about six months after release and for another six months.
● **Diet:** Piscivore.
● **Compatibility:** This species is quite capable of keeping fry numbers down in a community tank. Males are violently intolerant of other cichlids with a metallic blue courting dress and quite capable of assuming a

Below: **Chilotilapia rhoadesi** ♂ *A magnificent male.*

Above: **Sciaenochromis ahli** ♂ *A small but efficient piscivore.*

hyperdominant role if crowded. Does best in tanks more than 150cm (60in) long.

The fish begin spawning at 12 months, but it may take them several months more to attain full reproductive competence. Males court females very forcefully and are inclined to harass them after spawning is complete, so provide plenty of shelter in the breeding tank. Spawns can number 100 eggs, but 50-60 is the average. Fry tend to grow unevenly and are prone to sibling cannibalism. It is important to prevent such behaviour, for *S. ahli* tends to produce broods with a male-biased sex ratio of 3:1 to 4:1. This imbalance is worsened if the smaller females are devoured by their brothers.

This is one of the most spectacularly coloured cichlids to date exported from Lake Malawi. Because males do not attain their full coloration until they are one year old, adult specimens tend to be quite expensive. The nondescript juveniles are much more reasonably priced. Always try to buy a few of the smallest available fry to be certain of obtaining at least one female.

Cyrtocara moorii

Malawi blue dolphin

● **Habitat:** This species is one of the chisawasawa, or sandy bottom associated cichlids. Single large individuals or small numbers of juveniles live in close association with inshore foraging groups of sand-sifting cichlids, such as *Fossorochromis rostratus* or the various *Lethrinops* species.
● **Length:** Males up to 20cm (8in); females up to 16.5cm (6.5in).
● **Sexing:** Unlike most Malawian haplochromines, both sexes of *C. moorii* share the same coloration. However, males usually have a more pronounced hump than females. Their longer soft dorsal and anal fins are a more reliable indicator of sex.
● **Diet:** Micropredator.
● **Compatibility:** Despite its adult size, *C. moorii* is a rather unaggressive species that fares poorly when housed with larger or more belligerent tankmates. It does best in tanks more than 150cm (60in) long, in the company of smaller members of its own kind or with the various *Aulonocara* species. With them, *C. moorii* can establish the same close relationship it enjoys in the wild. Not apt to become hyperdominant.

Above: **Cyrtocara moorii** ♂♀ *Juvenile male (front) with female.*

Above: ♂ **Cyrtocara moorii** *Adult male's nuchal hump.*

Males do not always excavate a spawning pit, and courtship is rather low key. Pairs have difficulty coping with more active fish, so avoid introducing tankmates such as the various mbuna. Parental females do not usually show a prominent throat bulge, but they can be recognized by their conspicuous lateral spots and indistinct vertical barring. Spawns can number 100 eggs, but 60-80 is the average. The blue adult coloration begins to develop at 2.5cm (1in) and is fully developed by six months at 5cm (2in). Reproductive activity begins at about one year, when males are 7.5cm (3in) long.

Copadichromis quadrimaculatus

● **Habitat:** Off rocky shores in areas of local upwelling, known as chirundu, where it finds a plentiful supply of zooplankton. A pelagic species; one of the utaka group of specialized zooplankton feeders.
● **Length:** Males up to 18cm (7in); females slightly smaller.
● **Sexing:** Males begin to change colour at 6.5cm (2.5in), about six months after release. Full male coloration is not attained until 12 months after spawning, and the fish may require an additional two years to reach full adult size.
● **Diet:** Micropredator.
● **Compatibility:** As for *C. moorii*, although males are somewhat intolerant of other utaka, particularly if they sport a similar colour pattern. Uncomfortable with mbuna and too small to hold its own with the larger *Haplochromis*

Above: **Copadichromis quadrimaculatus** ♂ *A splendid courting male.*

species, *C. quadrimaculatus* does best in the company of small, bottom-living haplochromines, such as the peacocks of the genus *Aulonacara*. One of the few Malawian cichlids to appreciate a fairly deep aquarium.

Males either defend a flat surface as close to the water surface as possible and induce a female to spawn there in the usual haplochromine manner, or they orient their displays to a vertical surface. In the latter case, the pair's spawning behaviour resembles that of the various *Cyprichromis* species of Lake Tanganyika. Brood sizes range from 20 to 80 eggs, with clutches of 40-50 closer to the norm. Sexual maturity is attained at about nine months, at 9cm (3.5in) for males, slightly less for females.

At least a dozen other utaka have been imported, of which half are undescribed. Their modest adult size recommends them highly to aquarists lacking the tank space to house the larger Malawian haplochromines.

Below: **Protomelas similis** ♂ *Aptly known as the red empress.*

Above: **Nimbochromis venustus** ♂ *An attractive species.*

Protomolas similis

Red empress

● **Habitat:** Frequently found among *Vallisneria* beds.
● **Length:** Males up to 12.7cm (5in); females slightly smaller.
● **Sexing:** Males begin to change colour at 6.5cm (2.5in), about 6 months after spawning.
● **Diet:** Micropredator.
● **Compatibility:** As described for *C. moorii* (see page 148). Although males are somewhat more aggressive than that species, they seldom become hyperdominant in captivity.
● **Breeding:** Males are sexually mature at about one year. Females begin breeding at about the same time and spawns can number up to 100 fry.

At least three similarly coloured species have been marketed under the name red empress. The fact that none of them can be matched to a described species suggests that the *P. similis* group is in need of competent taxonomic attention!

Nimbochromis venustus

Venustus

● **Habitat:** Over inshore sandy substrates to depths of 10m (33ft).
● **Length:** Males up to 20cm (8in); females to 15cm (6in) in nature. Captive individuals are usually somewhat larger.
● **Sexing:** Males grow much more quickly and their colour metamorphosis begins at 10cm (4in), eight months after spawning.
● **Diet:** Piscivore.
● **Compatibility:** Like any predator, *N. venustus* will make a meal of any fish it can conveniently swallow. Conversely, it is not a good idea to house this species with other large Malawian piscivores; it is the smallest representative of the group and usually ends up on the losing side of aggressive encounters. Provide a tank at least 150cm (60in) long.
● **Breeding:** Spawns can number 200 eggs, but the usual range is 80-120. The fry grow rapidly but unevenly, so it may prove necessary to sort them to minimize losses from sibling cannibalism. The young may begin spawning at eight months, but such efforts are rarely successful. Full reproductive competence is typically attained between a year and 14 months after spawning.

This striking species is the smallest of a quintet of closely related benthic predators that consists of *N. polystigma*, *N. linni*, *N. fuscotaeniatus* and *N. livingstonii*. All can be managed in the same manner as *N. venustus*, but need larger tanks.

Central American cichlid species

In this part of the book we look in detail at a representative selection of 33 species of Central American cichlids, both established favourites and brilliant newcomers to the aquarium scene. They are presented in alphabetical order of scientific name, with the most widely used common names listed for each species. The majority fall within the genus *Cichlasoma*, and for simplicity we have selected to keep them there. As with so many branches of natural science, moves are continually underway to reclassify fish into different taxonomic groups. Such moves have become particularly strong as far as Central American cichlids of the genus *Cichlasoma* are concerned. Revisions of scientific nomenclature are the production of detailed study at the highest academic levels. Since the concerns uppermost in the minds of professional ichthyologists do not necessarily coincide with the practical considerations of amateur fishkeepers, however, we have

resisted the temptation to enter the nomenclature fray until a clear picture has emerged. In the interim, therefore, we have not introduced the names *Archocentrus, Parapetenia, Thorichthys,* and *Theraps* currently being suggested as possible subdivisions of the genus *Cichlasoma*.

That said, we have reflected the confusion that *does* exist within the *Cichlasoma* genus where it is relevant to fishkeepers and the identification and availability of aquarium stock. Such confusion often results from the supposed hybridization between species.

For each species featured in this section we provide details of its habitat, eventual size, diet, sex differences, and its compatability and breeding potential in the aquarium. Such descriptions, together with the colour photographs that accompany them, serve to underline the exciting potential that the Central American cichlids represent in the fishkeeping hobby worldwide.

Cichla ocellaris

Eye-spot Cichlid

● **Distribution:** Rivers and lakes of Central and South America.

● **Length:** Can grow up to 60-70cm (24-28in).

● **Sexing:** Juvenile fish cannot be sexed using visual characteristics. Even large fish do not show obvious differences, although males are thought to be more colourful.

● **Diet:** Juveniles (under 10cm/4in) are notoriously difficult to feed. Some exporters rely totally on live *Tubifex*. It may be possible to encourage small specimens to take earthworms and *Gammarus* shrimps. Large specimens (30cm/12in upwards) will normally starve unless offered live fish as prey, which may not appeal to many aquarists. Weaning large wild-caught *Cichla* away from taking live food is very difficult, and sometimes impossible.

● **Compatibility:** The Eye-spot Cichlid is a true fish predator and will undoubtedly consume any fish that will fit between its jaws. Despite this characteristic, *Cichla ocellaris* can be kept with large *Cichlasoma* species fairly easily. It does not appear to exhibit the same territorial aggression as *Cichlasoma* species, perhaps because of its pelagic, or mid-water, existence.

● **Breeding:** Commercial breeding may be practical, but few aquaria could accommodate a spawning pair. An unpublished report from Germany suggests that a cichlid enthusiast has achieved spawning success, with the pair showing typical parental care.

A long, laterally compressed cichlid well suited to a predatory life.

Its streamlined body shape enables it to put on impressive bursts of speed in pursuit of its fish prey.

Cichlasoma aureum

Gold Cichlid; Blue Red-top Cichlid

● **Distribution:** Southern Mexico and Guatemala.

● **Length:** 15cm (6in).

● **Sexing:** Males show dorsal and anal fin extensions, although this is much easier to see in mature pairs.

● **Diet:** *Cichlasoma aureum* is not difficult to accommodate because it will accept a wide range of prepared foods. It shows a preference for larval foods, such as bloodworm and gnat larvae. It will also take flake, foodsticks, frozen

Below: **Cichla ocellaris**
This juvenile specimen is the size and colour phase most likely to be encountered by fishkeepers. Small individuals are difficult to feed.

brine shrimps, *Mysis* shrimp and *Gammarus* shrimp.

● **Compatibility:** This species should be considered in the same mould as the Firemouth Cichlid (*Cichlasoma meeki*) because its body shape and size are almost identical. As a small to medium-sized cichlid, it is well suited to aquaria in the 90-120cm (36-48in) range.

● **Breeding:** A spawning pair show an intensification of colour, especially the male, which can appear in iridescent green. The pair clean and guard a chosen spawning site – usually a

Above: **Cichla ocellaris**
Adults such as this one, will thrive in a large cichlid community. This species prefers live food.

Below: **Cichlasoma aureum**
A beautiful cichlid, but one that is not widely available in fishkeeping circles. This fine male specimen shows the typical fin extensions.

flat surface of slate or bogwood. Once the eggs are produced and fertilized, the parents fan the batch of up to several hundred with a sweeping movement of their ventral fins. The viable eggs hatch in 48-72 hours, depending on the aquarium temperature. The brooding female displays a lighter ventral area to enable the fry to identify her once they are free swimming. The parents dig several pits in the substrate and move the fry from site to site during the first week or so.

The red-tipped dorsal fin of *Cichlasoma aureum* gives this medium-sized cichlid an appealing colour pattern. Unfortunately, the Gold Cichlid (or Blue Red-top Cichlid) as the author suggests its common name should be) is less widely available than some species.

Cichlasoma carpinte

Blue Texas Cichlid, Green Texas Cichlid

● **Distribution:** Mexico, River Panuco.
● **Length:** 30cm (12in).
● **Sexing:** Males develop a head hump, which increases in size as the fish reaches maturity. In breeding colours, males develop iridescent silvery blue markings on the body; females lighten in the ventral region.
● **Diet:** Shrimp, earthworms, insect larvae, foodsticks, pellets, flake food and, occasionally, leaf spinach.
● **Compatibility:** Dominant males can be destructive in a community aquarium, although they are ideal robust cichlids capable of holding their own among a group of large specimen fish (ie one of each number of species).
● **Breeding:** Commercially bred specimens are frequently available, but few aquarists are interested in spawning the species. A mature pair will choose a spawning site and defend the area in typical cichlid fashion. Large broods are reported, with improved growth and survival rates resulting from removing some of the developing fry from the parents during the first two weeks of them becoming free swimming.

The dark blue of this species is instantly recognizable from the pale blue-green of *C. cyanoguttatum.*

Below: **Cichlasoma carpinte**
This species can be considered one of the most adaptable and robust of Central American cichlids available to aquarists.

European fishkeepers know this form as the Texas Cichlid and the pale forma *C. cyanoguttatum* as the Cuban Cichlid. The names are frequently reversed in the USA, where the Texan populations of *C. cyanoguttatum* (see page 158) are proudly known as America's National Cichlid.

Cichlasoma citrinellum

Red Devil; Midas Cichlid; Lemon Cichlid

● **Distribution:** Crater Lakes of Nicaragua, The Rio San Juan system.
● **Length:** 25cm (10in).
● **Sexing:** Sexually mature males will develop a distinctive hump on the head – a so-called 'nuchal' hump.
● **Diet:** Shrimp, earthworms, spinach, foodsticks and large flake food.
● **Compatibility:** Aggressive when adult.
● **Breeding:** Large specimens often clash violently during pre-spawning activity. Jaw locking and chasing are permanent types of behaviour pattern during this period. In a group of cichlids the pair will usually become territorial and vigorously defend a corner of the aquarium from allcomers. If they are sexually mature and ready to spawn, the two fish will display breeding tubes – small points between the ventral and anal fins. Sometimes the pair will engage in extended cleaning of a rock or piece of bogwood. Alternatively, if the female is not ready, the male will attack her. This is shown by the female 'cowering' or being beaten into submission in the upper corner of the aquarium. In this situation, the female can suffer extensive body and fin damage. If the spawning is successful, several hundred eggs are produced. After 24-48 hours, the infertile eggs show signs of fungus attack; the parents will clean off the debris from unviable eggs, although sometimes the fungus spreads to developing eggs. Once the fry have emerged, the parents will often move them from site to site, picking each one off, one at a time. The fry of *C. citrinellum* are usually free swimming on the fifth or sixth day after spawning and will follow the parents around the spawning area as they search for food.

Below: **Cichlasoma citrinellum**
This large male displays the head hump (or 'nuchal' hump) that is characteristic of adult Red Devils.

A problem exists relating to the identity of *Cichlasoma citrinellum* and *Cichlasoma labiatum*. The latter is thought of as a large-lipped species, usually displaying some orange-red pigment. *C. citrinellum* is found in yellow or red forms and is a simple-lipped species. The problem is caused by the polymorphism described clearly by Dr Paul Loiselle in commercial literature (following Professor Barlow's work published in 'Investigations of Ichthyofauna of the Nicaraguan Lakes') as the *Cichlasoma labiatum* species complex. Scientists have observed introgressive hybridization between *C. citrinellum* and *C. labiatum* in the smaller lakes, where the lack of suitable partners makes such hybridization the only viable alternative for survival.

Aquarium stock has developed from imports into the USA and Germany during the 1960s and 1970s. Therefore, the original genetic strength of colour, size and shape have become diluted. Nonetheless, *C. citrinellum* has made an indelible mark on the aquarium hobby.

Cichlasoma cyanoguttatum

True Texas cichlid

● **Distribution:** Mexico, in Rio Grande, Rio Pecos, Rio Conchos.

● **Length:** 25cm (10in).

● **Sexing:** Males develop the frontal, or nuchal, hump and can be slightly stronger in colour.

● **Diet:** Shrimp, earthworms, insect larvae, foodstick, leaf spinach and flake food.

● **Compatibility:** Large individuals can become dominant in a cichlid group and will relentlessly bully subdominant or weaker fish. Spawning pairs undoubtedly require isolation from a community aquarium.

● **Breeding:** Spawning pairs are dark in the mid-body/caudal region, light from the head back to the centre of the body. The female prepares the spawning site while the male chases off intruders. The parents share responsibility for cleaning the eggs, although the female takes on the major part of the work. The female digs pits in which to place hatching fry and frequently moves them from site to site. Fry growth is dependent on the frequency and amount of feeding. Rearing a proportion of the fry separately ensures that a reasonable number develop.

Above:
Cichlasoma cyanoguttatum
The True Texas Cichlid is widely available and an ideal candidate for large cichlid community systems.

This species was confused in Europe under a false name, 'C. tetracanthus'. Its name has been long associated with the Blue Texas Cichlid (*C. carpinte*) and will continue to be so for many years because of the name confusion in commerical literature.

Cichlasoma dovii

Dow Cichlid; Dow's Cichlid; Wolf Cichlid

● **Distribution:** Lake Nicaragua, Costa Rica, Honduras.

● **Habitat:** In clear water areas near submerged or waterlogged brush or tree trunks.

● **Length:** 50-70cm (20-28in).
● **Sexing:** Males often display a more intense colour pattern and a larger, more pointed anal fin, although this can be considered only as a guideline to sexing in most cases.
● **Diet:** Aquarium-raised specimens will accept a wide range of shrimp, prawn and earthworm-sized foods. Large specimens will eat fish, so it is important to keep them with large cichlids.
● **Compatibility:** The Dow's Cichlid's attainable size puts this fish into the *superclass*, and excludes it from all but the largest aquarium. Few aquarists have kept this cichlid with other community cichlids; its comparative rarity solves the problem for most fishkeepers.
● **Breeding:** Aquarium spawnings are rare but not unheard of among cichlid enthusiasts. Large broods are reported, with typical parental care being observed.

Cichlasoma dovii has been sporadically available over the years and is well known among American Cichlid Association members through a large specimen named 'Pablo' kept by an enthusiast. It has been reported in popular USA magazines by Dr Paul Loiselle that the fish was as well known to the ACA membership as its owner. This shows that large individual cichlids have character and can become true pets.

It is interesting to note that the Dow Cichlid is a major predator of *Neetroplus* in Nicaraguan Lakes.

Some authors recommend a 200-1 ratio of fish to tank space to maintain an adult Dow Cichlid.

Below: **Cichlasoma dovii**
A pair of Dow Cichlids in typical spawning colours. The male (above) is clearly larger than the female.

Cichlasoma friedrichsthalii

Friedrichsthal's Cichlid

● **Distribution:** Widespread throughout Mexico, Guatamala, Honduras, Belize, Costa Rica and Nicaragua.

● **Length:** 25cm (10in).

● **Sexing:** Males are larger than females, display a more ornate body speckling and sport elongated dorsal and anal fin rays.

● **Diet:** Although a fish eater in its natural habitat, young farm- or tank-raised specimens will take almost any prepared food offered to them.

● **Compatibility:** As with all of the large *Cichlasoma* species, this fish is best kept with similarly sized cichlids, such as *C. synspilum* and *C, maculicauda*. Juveniles are similar in appearance to *C. motaguense* and are also reminiscent of *Petenia* youngsters.

● **Breeding:** Aquarium spawnings are rarely mentioned in literature, although it is reasonable to assume that this species spawns in much the same way as closely related forms, such as *C. motaguense.*

Above: **Cichlasoma hartwegi**
An adult spawning pair of Tail Bar Cichlids with fry. The male is the upper and noticeably larger of the pair. This species, although a relative newcomer to fishkeeping, has proved fairly easy to spawn, which should ensure its continuing availability around the world.

Below: **Cichlasoma friedrichsthalii**
A female specimen displaying the characteristic lines and blotches extending from the eye to the tail.

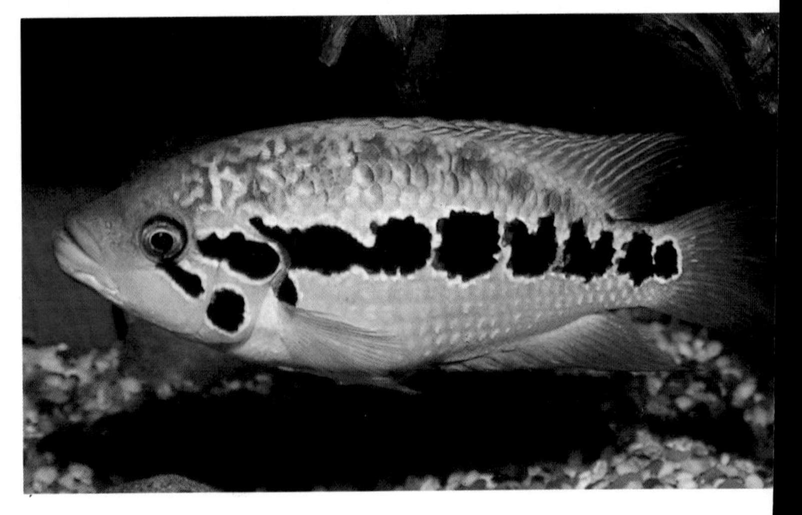

Cichlasoma friedrichsthalii is frequently confused with *Cichlasoma motaguense*, from which it can be distinguished by the vertical bars on either flank. A series of black blotches blend into this pattern between the base of the caudal fin and the eye. This part of the pattern is shared with *C. motaguense*.

Cichlasoma hartwegi

Tail Bar Cichlid

● **Distribution:** Mexico, in the Rio Grijalva.
● **Habitat:** Over sand and rock substrate.
● **Length:** 15cm (6in).
● **Sexing:** Males in spawning/brood-caring colour are silver with a hint of red speckling; the posterior half of the body has black criss-cross markings. The caudal base has a characteristic broad stripe (sometimes extending in an inverted arc from tail to eye), which the author refers to in the common name.
● **Diet:** Shrimp, insect larvae, leaf spinach and flake food.
● **Compatibility:** This is not known to be an aggressive species and would seem ideally suited to small to medium-sized community aquaria.
● **Breeding:** Successful spawnings have been recorded. It is said to be an easy species to breed, with typical parental care being shown by the breeding pairs. Moderately sized broods are produced; you can expect about 100-200 fry.

This species is similar to *Cichlasoma fenestratum*, which also comes from Mexico and grows to 20cm (8in), with pink to red fin edges and strong pink markings on the head. *C. fenestratum* and *C. hartwegi* have vertical body stripes, seemingly more pronounced in the former species, although the caudal peduncle bar is distinctive in *C. hartwegi*. The Tail Bar Cichlid was described in 1980, and so it is a relative newcomer to the list of Central American cichlids entering the hobby.

Above: **Cichlasoma intermedium**
A peaceful cichlid that has been made available to the hobby through the efforts of German importers and fishbreeders.

Cichlasoma intermedium

Intermedium Cichlid

● **Distribution:** Rivers in Mexico, Belize and Guatemala.
● **Length:** 15cm (6in).
● **Sexing:** Difficult to establish, although sexually mature males are more slender and can display fin extensions.
● **Diet:** Shrimps, earthworms, foodsticks and flake food.
● **Compatibility:** An excellent small to medium-sized species ideal for a Central American cichlid community aquarium.
● **Breeding:** Aquarium-raised specimens are not commonplace, although a reported spawning has resulted in this species becoming available in limited quantities. The spawning pattern is typical of closely related smaller species of *Cichlasoma.* The fry are said to develop well from the free-swimming stage on newly hatched brine shrimp; most fish fry will accept this live food.

This is a relatively new species in aquarium circles. The author encountered a specimen of *C. intermedium* for the first time in 1984, when aquarium-raised specimens became available from West Germany. The basic brown-green colour sometimes glistens as a border to the unique, reversed L-shaped body patterning.

Above right: **Cichlasoma labiatum**
A close-up showing the pronounced lips of an adult male – a distinctive characteristic of the species.

Cichlasoma labiatum

Large-lipped Cichlid; Red Devil

● **Distribution:** Lakes in Nicaragua.
● **Length:** 20-25cm (8-10in).
● **Sexing:** Males display pronounced lips and often have long, extended anal and dorsal fin rays.
● **Diet:** A great deal of research has been made into the feeding patterns of members of the *C. labiatum* group. In lakes and rivers, they are recorded as feeding on snails, organic debris, small fish, aquatic insects and fish eggs. In aquaria, this can be mirrored with frozen or freeze-dried foods.
● **Compatibility:** Apart from a continual desire to dig into the aquarium substrate, the Red Devil is as compatible within a Central American cichlid community as any comparably sized species.

● **Breeding:** Aquarium-raised specimens spawn easily, although parental care appears to diminish rapidly within the first week of the fry becoming free swimming. Removing the fry into a rearing tank is essential. Spawning males can prove to be extremely aggressive and excessive jaw locking, tail beating and biting can leave a female very much the worse for wear. Large fish can produce up to 7500 eggs!

Several colour forms exist, including white, yellow, yellow-orange and red. The large-lipped form with a red body is said to be extremely scarce in nature and correspondingly rare in aquarium circles. Juveniles are grey to green, with several cross bands. Hybridization has occurred between this species and *C. citrinellum*. (See page 157.)

Below: **Cichlasoma labiatum**
This colour form is one most likely to be seen by fishkeepers. Red-bodied specimens are rare.

Cichlasoma longimanus

Rose-breasted Cichlid

● **Distribution:** Mexico (Rio San Juan Basin), Guatamala, Honduras and Costa Rica, especially warm mud-bottomed backwaters.

● **Length:** 20cm (8in).

● **Sexing:** Females have a distinctive colour pattern in the dorsal region extending into the dorsal fin. (This characteristic has even led to the sexes being confused as different species by some hobbyists.) Adult males display distinctive lateral line blotches and possess longer dorsal and anal fins than the females.

● **Diet:** Crustaceans, insect larvae and prepared flake foods.

● **Compatibility:** This is not an aggressive species, although a breeding pair will spawn in the open on the substrate and, to quote Dr Paul Loiselle retelling a friend's experience with this fish, they will plonk the eggs down on a solid surface and dare any aquarium fish present to do something about it!

● **Breeding:** As a substrate sifter, the Rose-breasted Cichlid will choose a site on the gravel bed and dig out a pit. Up to 500 eggs are deposited, more or less in a pile. In the wild, they are said to choose sites within rocky areas. Males will spawn with several females, joining in brood

protection until the fry are free swimming, before entertaining the next female.

American stock orginated from Lake Jiloa in Nicaragua and Costa Rica, but the species has yet to become established in the hobby.

Cichlasoma maculicauda

Black Belt Cichlid

● **Distribution:** Southern Mexico, Guatemala, Belize, Costa Rica and Panama.

● **Length:** 30cm (12in).

● **Sexing:** Sexing juvenile Black Belt Cichlids can be achieved only with a certain amount of educated guesswork. In groups of semi-adult specimens (about 15cm/6in), the males exhibit early sexual behaviour as they establish a pecking order and display to the females. Adult males are slender in direct comparison to females, often displaying long anal and dorsal fin rays and showing stronger red in the caudal fin.

● **Diet:** Only tank-raised fish are available and these will feed on almost any aquarium prepared food. Mature fish will relish *Gammarus* shrimp and

Below: **Cichlasoma longimanus** *A pair of Rose-breasted Cichlids, (the male is the upper fish). An attractive newcomer to the hobby.*

large earthworms.
● **Compatibility:** This is a typical *Cichlasoma* species of medium to large size, once much sought after by aquarists. It is ideal for a general community system of cichlids. Its colour pattern of red cheek and caudal fin will enhance any Central American cichlid display.
● **Breeding:** Aquarium spawnings have been achieved, although not frequently, as in the case of *C. nicaraguense* and *C. synspilum*. Spawning males darken under the head and forward ventral area whereas females display a white ventral region after spawning. The distinctive colour change in the female helps to attract fry so that the brood can be controlled.

Juveniles rarely show the red patterning, although the distinctive vertical mid-body stripe and lateral caudal predruncle bar make them quite distinguishable from similar species, such as *C. synspilum*, which lacks the mid-body vertical bar pattern of *C. maculicauda*.

Below: **Cichlasoma maculicauda**
This male specimen displays the beautiful red patterning that ensures the popularity of this species among cichlid enthusiasts worldwide.

Cichlasoma managuense

Jaguar Cichlid; Managua Cichlid
● **Distribution:** Honduras, Nicaragua and Costa Rica.
● **Habitat:** In turbid slow-moving waters, especially in small tributaries of large rivers. Found close to rock substrate in Lake Nicaragua.
● **Length:** 30cm (12in).
● **Sexing:** Adult males appear more ornately patterned than females and usually display extended anal and dorsal fin rays.
● **Diet:** As the fish available are usually small tank-raised specimens, they will eat virtually any prepared foods, although it is not advisable to feed them totally on a diet of pellets or foodsticks as they will tend to produce a considerable amount of organic debris in the system. A diet of fish, shrimp and earthworms will ensure fast growth.
● **Compatibility:** Large specimens of this species can be particularly aggressive among their own kind and to small fish, but

Above: **Cichlasoma managuense**
An adult male Jaguar Cichlid in dominant colour form – the highlight of any large cichlid community.

youngsters are reasonably compatible with similarly sized fish in the aquarium.

● **Breeding:** Successful aquarium spawnings are recorded, usually at the expense of other aquarium occupants, which are relentlessly harassed by the parent fish in protection of their offspring.

The aquatic hobby confused this species with *Cichlasoma motaguense* for many years. However, the silver background colour in the Managua Cichlid (particularly bright in the colour form known as the Jaguar Cichlid) is not seen in the Motaguense Cichlid, which has a yellow-brown base to its overall colour patterning.

Cichlasoma meeki

Firemouth Cichlid

● **Distribution:** Rivers in Mexico (Yucatan) and Guatemala.

● **Length:** 15cm (6in).

● **Sexing:** Males are more colourful than females, have larger extensions to the dorsal and anal fins, and are often seen blowing out their throat and gill membranes (branchiostegals) in breeding displays.

● **Diet:** Bloodworms and shrimps are favourite foods for Firemouths, although they will accept a wide range of prepared foods in the aquarium.

● **Compatibility:** The Firemouth Cichlid is one of the first species form the Central American group to be encountered by fishkeepers. To aquarists used to the aggressive behaviour of larger *Cichlasoma,* this small species would appear a peaceful fish. In a general community aquarium, however, aquarists would consider the Firemouth Cichlid a bully; it is capable of consuming smaller tropical fish. It is suited to the smaller aquarium and, when adult, can be kept with larger Central American cichlids.

● **Breeding:** A spawning pair show an intensification of colour, with the male developing a particularly bright red throat and belly – hence the common name. Once an area has been selected, the male defends the territory, chasing off all other fish while he impresses the female with his 'peacock' style of display, which involves blowing out his gills and displaying his fins. The actual spawning site (usually a rock surface) is cleaned and protected by the pair, and then up to 500 eggs are produced by the female and fertilized at once by the male. Once hatched, their fry are kept in hollows dug out

by the pair until they are free swimming and capable of following the adults. They fry can be fed freshly hatched brine shrimp and powdered flake food. This, together with the cloud of chewed food the parents spit out after feeding, should see them safely through the 14-21 day stage, when they should be removed to a separate aquarium for growing on.

Firemouth Cichlids were first imported into the USA in about 1915 and into Europe in the 1930s. The present generation of aquarium fish are commercially bred in the Far East and the colour exaggerated by hormone feeding.

Cichlasoma melanurum

Black-blotch Cichlid

● **Distribution:** Lakes in Guatemala; rivers in Belize.
● **Length:** 20cm (8in).
● **Sexing:** Overall, females are less pigmented and have shorter fin rays than males. Adults pairs show a darkening of the ventral region.
● **Diet:** Crustaceans, insect larvae, leaf spinach and flake food.

Above: **Cichlasoma meeki**
A male Firemouth Cichlid, guaranteed to add striking colour to a community of small to medium-sized cichlids.

● **Compatibility:** This species is comparable to *C. maculicauda* and *C. synspilum* in its general temperament, and would be suited to community systems containing these species.
● **Breeding:** Very little has been published on the spawning of the Black-blotch Cichlid; they are reported to dig out gravel from under bogwood and spawn on the wood face. Typical parental care is shown; caring females are said to show intense colouring.

As a newcomer to the aquarium hobby, this species is not widely available. Its colour pattern is almost intermediate between *C. maculicauda* and *C. synspilum.*

Cichlasoma motaguense

Motagua Cichlid

● **Distribution:** Mexico, southern Guatemala (Rio Motagua), and El Salvador.
● **Habitat:** A noted river species.

Above: **Cichlasoma melanurum**
Similar in general appearance to the better-known Cichlasoma synspilum, *this medium-sized species has gained popularity over recent years.*

● **Length:** 30cm (12in).
● **Sexing:** Adult males are more ornate in colour pattern than females and sometimes exhibit slightly enlarged head fronts. Sexually mature females in breeding colours display indistinct vertical bands.
● **Diet:** Very much the typical large cichlid diet of whole shrimp, large chopped earthworms, foodsticks and pellet food.
● **Compatibility:** This *Cichlasoma* is certainly a boisterous cichlid when over half grown, but it is possible to keep them successfully in a large community of similarly sized cichlids.
● **Breeding:** Several spawnings have been recorded, and large broods have been successfully raised.

Cichlasoma motaguense has been confused with *C. managuense* by importers and for a period the two names have been interchanged. The two fish have a similar lateral blotch pattern, but *C. motaguense* has a yellowish body hue and retains the blotch line in the adult. By contrast, *C. managuense* displays the iridescent, almost reticulated, silver-black pattern.

The close similarity in patterns between *C. motaguense* and *C. friedrichsthalii* has also often led them to be confused in literature. A breeding female *C. motaguense* even shows the vertical bans distinctive in *C. friedrichsthalii*. Male gold forms are recorded in literature and, according to Dr Paul Loiselle, are known as 'El Rey de la Guapotes' – King of the Guapotes – in Nicaragua.

Below: **Cichlasoma motaguense**
A young specimen of this widely available species, showing the colour pattern and overall body shape most likely to be seen by fishkeepers.

Cichlasoma nicaraguense

Nicaragua Cichlid

● **Distribution:** Nicaragua (Lake Managua) and Costa Rica.

● **Length:** Males 25cm (10in); females 20cm (8in).

● **Sexing:** This species is fairly easily sexed in maturity; the females are slightly smaller than males and retain a simpler nonetheless bright coloration of yellow and blue-green. A lateral line black body stripe – prominent in juveniles – is retained by the female, but is seen as an indistinct central body spot in the predominantly yellow adult male.

● **Diet:** Juveniles and adults accept a wide variety of prepared and frozen food, and particularly relish chopped earthworms and fresh leaf spinach.

● **Compatibility:** This medium to large species is ideal for a large cichlid community aquarium.

It is not one of the most aggressive species of Central American *Cichlasoma* available to the hobby, but is still capable of causing considerable damage in a small fish community. Ideally, keep this species with *C. synspilum* and *C. maculicauda* until they sex out in size and colour, when surplus fish may be removed if necessary.

● **Breeding:** The Nicaragua Cichlid has proved easy to spawn once a compatible pair have formed a lasting bond.

A deep pit is dug in the substrate and the eggs placed into it in a clump. Fishkeepers have found it best to leave the parents to raise the fry, as this encourages future breeding success. In one instance, an aquarist removed the eggs to raise them artificially and subsequent noticed that the pair became aggressive towards one another.

Egg numbers depend on the maturity of the parents, but reports suggest a spawn of 300-500 eggs is not uncommon, with likely hatching rates between 20 and 50 per cent. Immature males will cause a greater number of eggs to be infertile. Therefore, early spawning can be expected to produce lower hatching rates. The eggs of this species are reported to be non-adhesive, a characteristic unique among Central American cichlids.

Above: **Cichlasoma nicaraguense**
A pair (the male above). These fish are extremely popular among fishkeepers for their brilliant colours.

This brightly coloured cichlid took European cichlid enthusiasts by storm when it first appeared in 1979/80. The colours on sexually mature pairs – brilliant greens, yellows and blues – rival those of many coral reef fish. The female displays a blaze of colour normally associated with outstanding males in some cichlid species. The superb adult colour pattern of the Nicaragua Cichlid has undoubtedly

caused a great upsurge of interest in keeping Central American cichlids. Juvenile fish are plain silver with black lateral stripe and mid-body blotch, and rarely herald the outstanding beauty of the adult.

Cichlasoma nigrofasciatum

Convict Cichlid; Zebra Cichlid

● **Distribution:** Guatemala, El Salvador, Honduras, Nicaragua and Costa Rica.

● **Length:** 10cm (4in).

● **Sexing:** Juvenile Convict Cichlids are difficult to separate into males and females, although the former tend to be more aggressively active when approaching sexual maturity. Also, adult males display particularly intense black vertical bands. These are less prominent in females, which – especially in breeding dress – are orange-yellow in the ventral region. And males invariably possess larger dorsal and anal fins that usually extend into filaments.

● **Diet:** The Convict must be the easiest species to accommodate on the dietary front; juveniles and adults with accept every form of tropical fish food. By enhancing the diet with bloodworm, chopped earthworms and frozen shrimp, a pair can be brought easily into spawning condition.

● **Compatibility:** Few community aquaria have not been plunged into chaos by the introduction of a renegade cichlid by the novice fishkeeper. Most aquarists will encounter cichlids for the first time in this manner and the most likely species to be purchased in ignorance of its aggressive tendencies is undoubtedly the Convict Cichlid. The extreme aggression stems from the cichlid's desire to spawn and protect its progeny. While it may infuriate the irate fishkeeper, who hates to see community fish victims of serious assault, this behaviour ensures that the fish will be successful in raising its fry, both in the natural habitat and in the aquarium.

● **Breeding:** This is one of the easiest species of *Cichlasoma* to spawn and raise. A breeding pair will choose a vertical of horizontal spawning site and defend the area vigorously from allcomers, including cichlids larger than themselves.

Commerically bred generations are said to be less protective of free-swimming fry and lose interest in the brood within a week, in some instances. It is wise to remove at least 50 per cent of the free-swimming fry to a separate raising system.

The zebra-like banding pattern is highlighted on the spawning male whereas the female develops a yellow underside to attract the brood.

Below: **Cichlasoma nigrofasciatum 'Kongo'**
The gold form of this species has become well established in the USA.

Wild-caught specimens, said to be highly coloured (especially in spawning conditions), are rarely encountered. The Convict Cichlid is commercially bred extensively in the Far East and also in the USA, where a gold variety is well established in the hobby.

Cichlasoma octofasciatum

Jack Dempsey Cichlid

● **Distribution:** Mexico (Yucatan), Guatemala and Honduras.

● **Length:** 20cm (8in).

● **Sexing:** Sexually mature males are blue-black in colour; females, however, tend to be paler in their markings.

● **Diet:** Commercial or tank-bred specimens will feed on a wide range of foods, although they show a preference for crustaceans, shrimps, snails, etc. Large specimens will greedily take foodsticks, whole shrimps and earthworms. This species will also show interest in green foods, such as leaf spinach or lettuce, but this should be offered only occasionally.

● **Compatibility:** Of the small to medium-sized cichlids, the Jack Dempsey Cichlid is probably the pugnacious, hence its common

Above:
Cichlasoma nigrofasciatum
An adult male, showing the normal colour form of the Convict Cichlid.

Below: **Cichlasoma octofasciatum**
This ebullient cichlid was one of the earliest species made available to fishkeepers. Striking blue spangling.

name. It is entirely suited to smaller cichlid communities if kept with Firemouth, Convict and Salvin's Cichlids. (See pages 167, 171, 177.)

● **Breeding:** Aquarium spawnings are said to have produced broods of up to 800 fry, although aquarium-raised breeding pairs are more likely to produce and raise much smaller broods. As the fry begin to show dark stripes on the body, remove them to a separate raising tank before the parent fish begin the spawning cycle again and lose interest in protecting the brood.

Known under the name of *Cichlasoma biocellatum* for some time in commercial literature, the Jack Dempsey is popular among newcomers to cichlids.

Above: **Cichlasoma panamense**
A juvenile, displaying the plain body colour reminiscent of Neetroplus nematopus. *Peaceful and easily bred.*

Cichlasoma panamense

Panama Cichlid

● **Distribution:** Small tributaries of the Chagres and Bayano Rivers in Panama.
● **Length:** 10cm (4in).
● **Sexing:** The adult male is larger than the female and, especially in breeding dress, displays a red body hue broken by a dark line of vertical blotches from the caudal base to the mid-point of the body. Brood-caring females display a pale ventral region and a broken vertical line of blotches on the body.
● **Diet:** A typical cichlid diet should enable the fishkeeper to develop the aquarium-raised specimens of this species that are available from time to time.
● **Compatibility:** Ideally suited to the smaller cichlid community system, this tenacious yet unaggressive species is one of the most recent newcomers to the European aquarist market.
● **Breeding:** Aquarium spawnings have been widely reported. Spawning pairs are said to prefer caves and plant pots as breeding sites. Brooding females display a light and dark body pattern, presumably for fry recognition and protection. This species was originally allied with *Neetroplus* and, from aquarium keeping aspects, can be considered similar to some extent. It appears that the Panama Cichlid does not share the awesome aggression of *Neetroplus* but is capable of producing and raising similarly sized broods in the aquarium.

Juveniles have a uniform grey body, which does little to recommend the species initially, but its small size and relatively peaceful nature will ensure its popularity as an ideal small community aquarium cichlid.

This species is a relative newcomer in comparison with other well-known forms; it became widely available for aquarium use in 1983 following a scientific survey carried out in the rivers of Panama.

Cichlasoma robertsoni

Metallic Green Cichlid

● **Distribution:** Lakes, lagoons and rivers in Mexico, Guatemala and Honduras.
● **Length:** 20cm (8in).
● **Sexing:** Males are larger and display a brighter coloration

Above: **Cichlasoma panamense**
An adult, illustrating the stark contrast in patterning when compared to the somewhat plain juvenile.

than females.

● **Diet:** Crustaceans, insect larvae and flake food.

● **Compatibility:** A moderate cichlid by normal *Cichlasoma* standards, this species is suitable for small to medium-sized aquarium communities.

● **Breeding:** Habitat observations in the wild reveal that this particular cichlid holds territories close to submerged tree trunks and large rocks in river conditions. It is capable of breeding in a cichlid community, even one containing *Neetroplus nematopus*, according to Dr Paul Loiselle.

The Metallic Green Cichlid is similar in shape and general appearance to the Firemouth Cichlid (*Cichlasoma meeki*), but differs in colour pattern and body spot. It has metallic green scales that catch the light, making it a most beautiful newcomer to the fishkeeping hobby. It is a substrate sifter and tends to dig continually into the aquarium gravel.

Below: **Cichlasoma robertsoni**
A handsome new cichlid, but not yet widely available to fishkeepers.

Cichlasoma sajica

Sajie's Cichlid

● **Distribution:** Small streams and lakes in Costa Rica.

● **Length:** 12.5cm (5in).

● **Sexing:** Adult males possess extended dorsal and anal fins and show a blue tinge to the body, whereas females are smaller and yellowish. Brooding females darken and display a thin vertical body stripe from the mid-dorsal region.

● **Diet:** A wide variety of foods will keep this small cichlid in good condition in the aquarium.

● **Compatibility:** It is ideal for a 60-90cm (24-36in) cichlid community aquarium, although sadly it is less widely available than the cichlid species most confused with it, *C.spilurum*. Sexually active pairs of *C.sajica* can be extremely unsociable towards other aquarium occupants and should be removed to a special breeding tank. If kept alongside larger cichlids, they will act as small food scavengers and will be rarely threatened by their larger cousins in the system.

● **Breeding:** This species is known as an open spawner, although observations show that it will take cover in pots or rock caves.

Cichlasoma sajica could easily be confused with *C. spilurum* and *C. spinosissimum*, although the red

Below: **Cichlosoma sajica**
A male specimen showing the clear similarity in shape of this rarely encountered cichlid to the widely obtainable Cichlasoma spilurum.

speckling in the finnage of sexually mature males would distinguish it fairly easily. (Also, the central vertical bar is usually prominent in juveniles and adults and is a characteristic of the species.) They are also fairly distinct in their geographical distribution: *C. sajica* is found in Costa Rica; *C. spilurum* in Guatemala and Belize; and *C. spinosissimum* in the highland rivers of Guatemala.

Cichlasoma salvini

Salvin's Cichlid; Tricolor Cichlid

● **Distribution:** Mexico, Guatemala, Honduras.

● **Length:** 15cm (6in).

● **Sexing:** Sexually mature males only two-thirds grown develop a red blotch in the ventral body/anal fin region, which makes this one of the most attractive of the smaller cichlids.

● **Diet:** This cichlid shows a preference for larval or shrimp foods and will retain its good colour if fed regularly on bloodworm and *Gammarus* shrimp.

● **Compatibility:** A tenacious dwarf Central American cichlid, Salvin's Cichlid will enhance any community, large or small. In

Above: **Cichlasoma salvini**
A striking female specimen that shows clearly why the beautiful Tricolor Cichlid is one of the most popular species available.

certain community aquaria, however, this beautiful cichlid would be bullied because of its distinctive pattern and smaller size. Breeding pairs can be very aggressive and will bite other aquarium cichlids viciously unless given an aquarium of their own for breeding activities.

● **Breeding:** This species is said to prefer placing eggs on a sloping or vertical surface. Egg numbers have been recorded at 1000, although half that figure is the accepted norm. Both parents show bright colour patterns, especially while they are protecting the eggs and fry. Other fish are said to associate the bright colour with aggression and endeavour to avoid them during this period. Fry hatch within three days and will soon take newly hatched brine shrimp and powdered food. Experience shows that it is best to remove the fry from the parents. It is also beneficial to separate the parent fish to save the female from the somewhat over-enthusiastic attentions of the male in his desire to repeat the spawning cycle *before* the female is ready to start breeding again.

The Tricolor Cichlid is not widely known in fishkeeping circles, but it is an ideal species to consider for a breeding programme. It is well suited to small to medium-sized aquaria because it reaches sexual maturity at the relatively small size of 10cm (4in).

Cichlasoma sieboldii

Bandit Cichlid

● **Distribution:** Pacific slopes of Costa Rica and western Panama.
● **Habitat:** Fast-flowing or still waters.
● **Length:** 10-15cm (4-6in) in aquarium specimens, although the maximum length recorded in literature is 30cm (12in).
● **Sexing:** Females are noticeably smaller than males.
● **Diet:** Crustaceans, insect larvae, vegetables, lettuce, leaf spinach and flake food. In the wild, it feeds by combing larvae from rocks.
● **Compatibility:** This can be an aggressive species, which suggests that it may be best kept with medium-sized *Cichlasoma* species.
● **Breeding:** Breeding pairs develop brooding colours that give the fish its common name. Lines across the forehead extend around the eyes to give the appearance of a fish wearing a mask. The throat turns dark grey, this colour being carried into the body through about seven coalescing blotches that extend from the pectoral fins to the tail. *C. sieboldii* is more or less an open spawner, although it will take advantage of bogwood or rock areas on which to lay eggs.

Cichlasoma sieboldii undergoes almost a 'Jekyll and Hyde' transformation when breeding by changing both its body colour – normally sandy overall with small red spots – and personality.

Cichlasoma spilurum

Blue-eyed Cichlid; Jade-eyed Cichlid

● **Distribution:** Guatemala and Belize.
● **Length:** 12.5cm (5in).
● **Sexing:** Males display a

Above: **Cichlasoma spilurum**
A male specimen of the Jade-eye Cichlid. This is an excellent species to introduce fishkeepers into the fascinating world of Central American cichlids.

yellowish ventral region and are generally longer in the body and in the dorsal and anal fins than females. Brood-caring females have a distinctive series of vertical black bars, not dissimilar to those of the Convict Cichlid, (*C. nigrofasciatum*).

● **Diet:** This species will thrive on much the same fare as other small to medium-sized cichlids, and will particularly relish bloodworm and *Daphnia*.

● **Compatibility:** Spawning couples – in keeping with the tenacious smaller cichlids – can be difficult; otherwise they are perfectly suited to living in small systems.

● **Breeding:** Breeding Jade-eyed Cichlids is relatively easy. Mature pairs will accept an upturned plant pot or rock cluster as a spawning site. Large broods can be expected and free-swimming fry should be removed to separate quarters when they are 14 days old.

Below: **Cichlasoma sieboldii**
The Bandit Cichlid has recently become available to fishkeepers and is ideal for small aquaria.

The Jade-eyed Cichlid is probably the most underrated of the small Central American species. A spawning pair display warm colours and contrasting patterns that merit greater interest being shown in this species. They prove excellent parents and will successfully raise broods within a community aquarium without too much disruptive behaviour.

Cichlasoma spinosissimum

False Blue-eyed Cichlid; False Jade-eyed Cichlid

● **Distribution:** Rio Polochic in the highlands of Guatemala.
● **Length:** 12.5cm (5in).
● **Sexing:** Adults males display a blue iridescence and have longer anal and dorsal fin rays than females.
● **Diet:** Crustaceans, insect larvae and most types of prepared foods.
● **Compatibility:** This species is ideally suited to small cichlid community systems.
● **Breeding:** Care-brooding pairs display vertical bands or stripes similar to those of the Convict Cichlid, (*Cichlasoma nigrofasciatum*), and the female darkens considerably in the ventral region. The spawning sequence is identical to that of *C. spilurum*, the cichlid it closely resembles.

This species is rather obscure in aquarium circles and probably confused in commerical literature. It may become available for the general aquarium market through the efforts of American or German collectors.

Above:
Cichlasoma spinosissimum
This species is thought to be a colour morph of the similar C. spilurum.

Below:
Cichlasoma spinosissimum
When available, treat in the same manner as Cichlasoma spilurum.

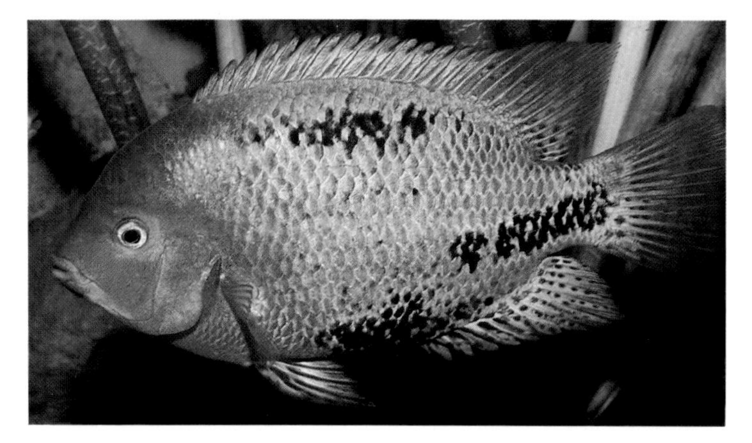

Above: **Cichlasoma synspilum**
A female specimen, illustrating how vibrant this widely available species can be. Males display nuchal humps and an even brighter colour.

Cichlasoma synspilum

Firehead Cichlid; Quetzal

● **Distribution:** Guatemala and Belize.
● **Length:** 30cm (12in).
● **Sexing:** Adult males are brighter in colour than females and develop a slight nuchal hump. Juveniles can be sexed at the half-grown stage using the developing males' brighter colour pattern and fin development as a guide.
● **Diet:** Young specimens will take almost any prepared food. Adults can become fussy, preferring large prawns, pellets, leaf spinach and earthworms.
● **Compatibility:** This brightly coloured species will enhance any large Central American cichlid community. Together with *C. nigaraguense, C. synspilum* must take a good deal of the credit for promoting Central American cichlids to the aquarium hobby. If kept with similarly sized species, Firehead Cichlids will co-exist peacefully, although once a sexually mature pair have formed a bond, greater aggressive behaviour can be expected.
● **Breeding:** A spawning pair can produce large broods, although the difficulty lies in finding two compatible specimens. Pair bonding can result in continual spawning cycles. Fry rearing is best undertaken by removing 50-75 per cent of the brood to a separate aquarium. If some fry are allowed to progress with the parents, there will be a high mortality rate in the period 7-14 days after hatching, but continued parental care and pair bonding will be ensured by taking such action.

Cichlasoma synspilum is one of the most widespread of the commercially known larger species and is the species most likely to be available at your local dealer. To develop a group of *C. synspilum* with the priority of creating a good spawning pair, start by buying 4-6 juveniles. If possible, buy half of these from a different source to prevent close inbreeding. Raise the group in a 120 × 45 × 30cm (48 × 18 × 12in) aquarium or similarly sized system. As the cichlids become sexually active, the dominant male will display to the best female. At this stage, take the balance of the subdominant fish back to the retailer as surplus to requirements. Although you will lose a good percentage of the original purchase price, you will have secured a compatible pair that should produce reasonable broods for you to raise and sell on later.

Cichlasoma trimaculatum

Shoulder-spot Cichlid; Three-spot Cichlid

● **Distribution:** Lagoons, lakes and rivers in Mexico, Guatemala and El Salvador.
● **Length:** 30-35cm (12-14in).
● **Sexing:** Males have extended finnage and display a stronger spot pattern than females. Males are larger and deeper in the body and generally more colourful than females, with a claret-coloured throat.

Above: **Cichlasoma trimaculatum**
A wild-caught male displaying the distinctive colour patterns reflected in its common name.

● **Diet:** Large shrimp, earthworms, foodsticks and pellet food.
● **Compatibility:** As a large-sized species, the Shoulder-spot Cichlid should be considered

Below: **Cichlasoma trimaculatum**
A spawning pair in typical breeding coloration. The male is the upper fish. Aggressive when breeding.

suitable only for large aquaria containing similarly sized cichlids.
● **Breeding:** As an open spawner, the parent fish select and defend their spawning site with a great deal of aggression.

This species has become widely available in recent years and earns its common name from the distinctive pattern of three body blotches, one of which is situated in the region of the shoulder.

Cichlasoma umbriferum

Blue Speckled Cichlid
● **Distribution:** Rivers in Panama and Colombia.
● **Length:** 30-40cm (12-16in).
● **Sexing:** Females are clearly smaller and less brightly coloured than males.
● **Diet:** Crustaceans, fish, large earthworms, pellets and foodsticks.
● **Compatibility:** This species can be considered as comparible and safe only if it is kept with cichlids of its own size.

Below: **Cichlasoma umbriferum**
A sexually mature male of good colour. Provide plenty of swimming space in the aquarium for this particular species.

● **Breeding:** This species is said to be an open spawner, although few reports exist of aquarium spawnings.

It is known that the Blue Speckled Cichlid is an open-water swimmer and thus requires a spacious aquarium. The bright blue spots on the greyish yellow body are distinctive to the species. Two black spots – one central on the body, the other on the tail base – help to identify juvenile fish.

Cichlasoma urophthalmus

Mexican Tail-spot Cichlid
● **Distribution:** Lagoons in Mexico (Yucatan), Guatemala, Belize and Honduras.
● **Length:** 20cm (8in).
● **Sexing:** Males display extended fins and grow larger than females.
● **Diet:** Shrimp, earthworms, foodsticks and pellets.
● **Compatibility:** Contrary to some descriptions, the author has found this particular species to be extremely aggressive. An adult specimen chased and attacked a larger Oscar (*Astronotus ocellatus*), until the latter had to be removed. for its own safety.

This particular incident occurred in a 240 × 60 × 60cm (96 × 24 × 24in) aquarium containing a selection of of cichlids.

● **Breeding:** Breeding pairs prefer to spawn under cover of bogwood or rockwork. Parental care is typical of these cichlids, although *Cichlasoma urophthalmus* can be said to be especially aggressive towards intruders near the spawning site.

C. urophthalmus is undoubtedly confused with *C. festae*, a South American species that shares the same red basic colour and caudal base spot, although this is larger in *C. urophthalmus, Cichlasoma festae* attains a greater body length and is more robust in overall appearance.

Herotilapia multispinosa

Rainbow cichlid

● **Distribution:** Lakes, streams and rivers in Nicaragua and Costa Rica.

● **Length:** 12.5cm (5in).

● **Sexing:** Male Rainbow Cichlids are more brightly coloured than females, with areas of yellow, brown and red overlaid with a line of black blotches that begins just behind the eye and extends along the mid lateral line. This is punctuated by a large spot or

Above: **Cichlasoma urophthalmus**
A male Mexican Tail-spot Cichlid. This species is often confused with C. testae *from South America.*

blotch just beyond the middle of the body and finishes with a blotch at the base of the tail. The female shares the same basic pattern, although drab by comparison, and is shorter overall with shorter fins.

● **Diet:** Young specimens will pick at any food which falls to the substrate and show a preference for larval foods, such as bloodworm and gnat larvae, which would be a natural food source in the wild.

● **Compatibility:** This is perhaps the most compatible of the small Central American cichlids available to aquarists. It is an excellent small dither fish among large *Cichlasoma* species, which will not feel threatened by its diminutive stature in the aquarium.

● **Breeding:** This species is easy to breed and will spawn when only halfway to adult size! Breeding females change from their usual drab pattern to take on a yellow hue, which appears to be for brood recognition. A pair will accept almost any site in the aquarium for spawning, including bogwood, rocks or plant pots. Aquarium-bred specimens are so far removed

Above: **Herotilapia multispinosa**
The Rainbow Cichlid is an ideal species to keep in a small community system. Very easy to breed.

from the original wild form that the colour pattern is reduced to a brown body with lateral banding or spots.

Neetroplus nematopus

Pygmy Green-eyed Cichlid

● **Distribution:** Lakes and rivers in Nicaragua and Costa Rica.
● **Habitat:** River populations live in fast-flowing waters above rock-strewn substrates.
● **Length:** 7.5-10cm (3-4in).
● **Sexing:** Adult males appear to develop a slight nuchal hump and usually can be identified by their dominant behaviour. Some aquarists suggest that males exhibit slight finnage extensions in comparison with females and this can be a useful guide.
● **Diet:** Lake populations are known to be algae grazers and should be offered fine leaf spinach, lettuce or peas. In addition, *Neetroplus* will accept any larval or shrimp food and will thrive even if fed solely on flake food. However, a varied diet soon brings sexually mature fish into breeding condition.

Below: **Neetroplus nematopus**
One of the smallest yet one of the most aggressive species available to the hobbyist.

● **Compatibility:** All aquarium rules are broken by this tiny Central American cichlid. A 7.5-10cm (3-4in) male can cause havoc in small and large aquaria alike, and it will not show any fear of large fish, especially if paired and brood protecting. Although ideally suited to small aquaria, these demons would not come unstuck in a busy large cichlid community!
● **Breeding:** These are cave- or hole-spawning cichlids. In the aquarium, a breeding pair will accept a small plant pot buried in the gravel (leaving an opening only large enough for them to squeeze through) or a narrow space in a rockwork cluster. Once the eggs have been produced, the female reverses her colour pattern. The body darkens from grey to black, and the characteristic black bar becomes white. (This colour change is one of the most extreme to occur among cichlids.) The male also reverses his colour pattern when brood protection begins. Although up to 100 eggs can be produced in one spawning, about 30-40 is the usual number. Brood-protecting parents will not hesitate to attack intruders, but other community fish tend to recognize the demon pair as potential trouble to be avoided.

Neetroplus nematopus is noted as a 'cleaner' fish to larger cichlids, much in the same way as the Cleaner Wrasse (*Labroides dimidiatus*) attends to its fellow marine fish.

Petenia splendida

Bay Snook; Giant Cichlid
● **Distribution:** Mexico, Guatemala, Belize and Nicaragua.
● **Habitat:** It lives in still waters, such as quiet stretches of large rivers, typically in densely planted areas near waterlogged brush or tree trunks.
● **Length:** 50cm (20in).
● **Sexing:** Males tend to display longer fin rays and are invariably brighter in colour than the females.

Above: **Neetroplus nematopus**
A male displaying the fin extensions and slight nuchal hump characteristic of the species. These fish are said to pick parasites off Cichlasoma nicaraguense *in the wild.*

● **Diet:** These are avid fish eaters, although this appetite for eating fellow aquarium occupants can be tempered by offering large earthworms, whole shrimp or prawns, maggots and fish pieces as a diet.
● **Compatibility:** A full-sized specimen would hardly suit a small community aquarium, although youngsters can be brought up easily among smaller Central American cichlids. In the semi-adult stage, it is ideally suited to a large aquarium system and can be kept with adult *Cichlasoma* and *Cichla* species.
● **Breeding:** Aquarium spawnings have been reported, although this is a rare occurrence compared with the many *Cichlasoma* species that are regularly spawned. As one of the giant – or true Guapote – species, *Petenia splendida* is capable of producing up to 5000 eggs in a spawning, although less than half of these will be fertilized. Large rocks or large plant pots (split down the middle and positioned vertically) should be placed in the breeding tank as potential spawning sites.

Juvenile Giant Cichlids had been available in recent years, although never in any regular quantity. Colours include red-eyed brown, black speckled, silver-black speckled, and red.

Above and below: **Petenia splendida** *The Bay Snook in adult colour (above) is a most attractive giant species. Juveniles (below) are more likely to be encountered.*

Nomenclature and species groupings

The taxonomy of South American cichlids is currently in a state of flux. The Swedish ichthyologist, Sven O. Kullander, began a re-evaluation of the names and placement of cichlids of the Americas in the early 1980s. His studies, to date, have concentrated on species from the Peruvian Amazon to the Guianas, but have produced results whose nomenclatural ripples have been felt throughout the previously accepted systematics of the neotropical *Cichlidae*. His work is currently unfinished and has not yet been embraced by all the ichthyological community. Nevertheless, many of the conclusions he has reached seem appropriate and useful and are therefore used here.

Why are scientific names important? Why should hobbyists care? Can't we get along with common names? There are several answers to these questions. Scientific names allow scientists and others to talk to each other about particular fish with precision. The latinized scientific binomial name is unique to each species and serves as a symbolic shorthand for that fish. Common names could do the same, but a standardized list of unique common names for each species does not yet exist. Many common names are used for several different fish. For example, there

are at least two cichlids with the common name '*Flag Cichlid*' in the aquarium hobby, *Laetacara curviceps* and *Mesonauta festivus*, but neither looks even remotely like the other nor are they closely related. So the latinized scientific names are the only way to identify what you actually have.

A second reason why scientific names are useful is that they attempt to express something about evolutionary relationships. For example, most aquarists would be easily persuaded that the many elongate, torpedo-like pike cichlids are closely related and may have arisen from a common ancestral piscivorous cichlid which had its basic body plan. Placement of the majority of pike cichlids in the genus *Crenicichla* attests to this relationship. Some of the pike cichlids have a reduced snout and consistent changes in the arrangement of their teeth. These have been placed in the genus *Batrachops.* Such placement suggests that these species are more like each other and form a natural sub-grouping within the greater set. Ichthyologists may debate that placement or suggest that the changes are too unimportant to warrant splitting the subgroup from the main assemblage (in fact, Kullander has lumped *Batrachops* together with *Crenicichla!*). Nevertheless, it is both interesting and important to

Above: **Dicrossus maculatus** *is a species which has been newly re-imported from the Brazilian Amazon.*

realize that these cichlids have solved similar ecological challenges in similar ways, and are presumably derived from a common ancestor. One can make too much of the evolutionary (phylogenetic) implications of pigeon-holing like species in the same genus. However, often natural groupings do speak volumes about the evolutionary processes of radiation and speciation. In a more practical sense, species groupings suggest commonalities in aquarium care and spawning to aquarium hobbyists attempting to maintain and captively breed fish. Many of the 'species groups' make intuitive sense to aquarists who have actually worked with these fish in the aquarium, rather than simply studying pickled examples.

For these reasons, I believe that Kullander is on the right track with his re-evaluation of the neotropical *Cichlidae*. Thus, the scientific nomenclature used throughout this species catalogue reflects the current state of this re-evaluation. I have tried to explain the reasons for the rearrangements to avoid undue confusion, and have tried to indicate the 'pre-Kullander' names under which information about these fish can be accessed through earlier aquarium literature. I have also presented the species in what I believe are their natural groupings as aquarium fish which, more often than not, reflect their actual evolutionary links. These major groups include the Acaras, the Eartheaters, the Apistos and allied dwarf cichlids, the pike cichlids, the Cichlasomines and derivatives, and finally, a few

THE ACARAS

This group of South American cichlids is the most primitive in terms of generalized body structure and ecology. Acaras are typically egg-shaped omnivores which, with certain exceptions, are monogamous, biparental substrate spawners. In the aquarium, they are amongst the least demanding of the South American *Cichlidae* with regards to water chemistry and quality or dietary requirements. They are also the least likely to enter the hobby as anything other than contaminants and are usually sold under the catch-all name 'Port Cichlid' or simply 'Acara'. In fact, this assemblage is rather specious and heterogeneous.

Heckel, in 1840, created the genus *Acara* for these fish, choosing the native Guarani name for these cichlids. Eigenmann and Bray, in 1894, put forward the genus *Aequidens* as a replacement for *Acara* when they discovered that the type specimen, *Acara crassipinnis*, was actually a junior synonym of the oscar, *Astronotus ocellatus*, a fish substantially different from most other Acaras. The name *Aequidens*, which translates as 'equal tooth', refers to the absence of enlarged pseudocanine teeth which are often found in members of the Cichlasomine lineage. All members of the genus *Aequidens* have instead, small conical teeth and most, but not all, have only three hard rays in their anal fins compared with four or more rays in the 'Cichlasomines'. Most of the Acaras have been written about in the hobby literate under the name *Aequidens*.

In fact, the genus *Aequidens* is a mixed collection of fish, Kullander has redefined the genus *Aequidens* and has restricted it to the larger forms of Acaras. These we will call the 'true' Acaras. He has also erected the genera *Bujurquina, Laetacara, Krobia, Cleithracara,* and *Guinacara* to hold most of the other 'orphaned' Acaras.

Those that he has not yet dealt with and which have not yet been formally reassigned are designated for now simply '*Aequidens*', in quotation marks.

TRUE ACARAS

The species roster of True Acaras (*Aequidens*) includes *chimantanus, diadema, metae, pallidus, paloemeuensis, patricki, plagiozonatus, potaroensis, tetramerus, tubicen, unicellatus, viridis.* Note that the aquarium fish *Aeq. awani* is in fact, *Aeq. viridis.* Some of these species, particularly *Aeq. tetramerus*, exist in several coloration/geographical varieties. Only a few of these species have been imported, most as accidental 'contaminants'. Most of these species grow quite large, about 25-30cm (10-12in), and are somewhat belligerent. All but *Aeq. diadema*, a primitive mouthbrooder, are biparental substrate spawners. Despite their belligerence, most are easily cared for.

miscellaneous species which seem to fit nowhere and which are probably transitional in an evolutionary sense. In certain cases, assignment of an aquarium fish to a definite species is difficult and tentative; many may well be new to science and undescribed. These fish are referred to as 'sp. affin.', short for *species affinis* which in scientific circles means 'looks like, has affinity with, a species of this name but we cannot be sure'. There are several entries in this catalogue bearing that designation. My coverage in this cataloge is nowhere near exhaustive, which is impossible for a book of this scope. However, species which can be expected to be encountered in the aquarium trade have been chosen as representative of their particular species group.

Aequidens metae

Rio Meta Acara

● **Distribution:** Rio Meta, Colombia.

● **Length:** 20-30mm (8-12in) in captivity.

● **Diet:** Undemanding type of omnivore.

● **Sexing:** Essentially isomorphic. Males slight more elongate than females.

● **Aquarium maintenance and breeding:** Although beautiful, *Aeq. metae* can be quite belligerent. Best suited for the mixed, 'rough' cichlid community. Undemanding with respect to water chemistry and quality, as well as feeding. Although growing quite large, these are precocial spawners which will breed at 10cm (4in) or less. At this size they handle easier. Typical biparental substrate spawners.

Aeq. metae is relatively uncommon in the hobby. It was originally described from the Rio Meta, a tributary of the Rio-Orinoco, and is found as an occasional juvenile accidental contaminant in cichlid shipments from Colombia. More recently, it has also been available in the form of tank-raised juveniles from Europe.

Above: **Aequidens metae** *is an attractive 'True Acara' from Colombia.*

Below: **Aequidens diadema** *is the only known mouthbrooding 'True Acara'.*

BLUE ACARAS
The Blue Acaras have yet to be generically reassigned, hence the use of quotation marks around the old genus name. This group of medium- to large-sized Acaras contains some of the more beautiful members of the Acara lineage. Species include *'Aeq.' biseriatus, coeruleopunctatus, latifrons, pulcher, sapayenis, rivulatus* and the *'rivulatus'* complex. Several of these, *Aeq. pulcher* and *Aeq. sp. affin. rivulatus*, have become staples of the aquarium trade and are commercially propagated in Asia and Florida.

'Aequidens' pulcher

Blue Acara
● **Distribution:** Northwestern South America, Trinidad.
● **Length:** 15-20cm (6-8in) in the aquarium.
● **Diet:** Undemanding omnivore.
● **Sexing:** Essentially isomorphic. Males slightly more elongate than females.
● **Aquarium maintenance and breeding:** Relatively undemanding with respect to water chemistry, quality and overall maintenance. They can be somewhat belligerent, so adequate shelter and appropriate tankmates should be provided. They are ready biparental substrate spawners, and will lay 100-500 eggs on rocks or other hard substrates, and make exemplary parents, even at 5-7cm (2-3in) size.

'Aequidens' pulcher hails from northwestern South America, the coastal regions of Venezuela including the island of Trinidad, down to the Orinoco drainage. However, since it has been commercially bred in Asia and Florida we rarely see wild specimens of this beautiful fish. '*Aeqidens*' *pulcher* is replaced by the closely related *'Aeq.' latifrons* in northern Colombia, a higher-bodied form with more iridescent scalation. *'Aequdiens' coeruleopunctatus* replaces *'Aeq.' pulcher* at the Colombian-Panamanian border and is found as far north as southern Costa Rica in Central America. Care is identical for all. Both *'Aeq.' pulcher* and *'Aeq.' coeruleopunctatus* have proved to be movable platform spawners in the wild habitat, using waterlogged leaves as the preferred egg receptacle.

Below: **'Aequidens' pulcher,** *the Blue Acara, is an undemanding mid-sized Acara.*

Above: *This male Green Terror,* **'Aequidens' sp. affin. rivulatus,** *lives up to its common name.*

'Aequidens' sp. affin. rivulatus

Green Terror, Rivulatus

● **Distribution:** Eastern Ecuador.
● **Length:** Males can exceed 30cm (12in) in captivity, females up to about 20cm (8in).
● **Diet:** Omnivorous.
● **Sexing:** Mature males develop conspicuous nuchal humps. Females stay one-third smaller and are not as elongate.
● **Aquarium maintenance and breeding:** A fairly belligerent fish that is appropriate in mixed communities of rowdy cichlids. Relatively undemanding as to water and dietary requirements. These are precocial spawners who reach sexual maturity at 10-15cm (4-6in). At this size, they are much less liable to liquidate each other in the process of establishing a pair bond. Large specimens can be difficult to pair. Reproduction as for most biparental spawning cichlids.

The real *'Aequidens' rivulatus* hails from the Pacific coast of Ecuador, Venezuela and Colombia, and is rarely imported. The hobby *'rivulatus'*, the Green Terror, was first imported in the early 1970s along with *'Cichlasoma' festae*, the 'Red Terror' from eastern Ecuador. In the real *'Aeq.' rivulatus*, the centres of the flank scales are iridescent green, whereas in the Green Terror, it is the edges of the scales, leaving the centres dark. *'Aequidens' sp. affin. rivulatus* may, in fact, be either *'Aeq.' azurifer,* or *'Aeq.' aequinoctalis*, two species earlier synonymized with *'Aeq.' rivulatus.* Two colour varieties of Green Terror are known: those with white edging to their tail and unpaired fins, and those with orange or red. They are apparently the same species. There may be several dwarf species of the *rivulatus* – complex, that grow up to 13cm (5in).

SMILING ACARAS

The Smiling Acaras are a group of mostly smaller forms which share a peculiar snout marking. The genus name, *Laetacara*, derives from the Latin *laetus,* meaning 'happy'. The 'smile' consists of a series of three dark stripes extending from the eyes to the lips. The species include *L. curviceps, dorsigera, flavilabrus,* and *thayeri.* With the exception of *L. thayeri*, the giant of the group at 15cm (6in), the Smiling Acaras remain small at 7.5-10cm (3-4in) and are perfect for the planted community tank. All are biparental substrate spawners.

Laetacara curviceps

Curviceps, Flag Cichlid

● **Distribution:** Amazon drainage.
● **Length:** Males up to 7.5cm (3in), females 1cm (½in) smaller.
● **Diet:** Requires frozen or live foods for optimal conditioning.
● **Sexing:** Males develop a convex head profile. In some populations, females may have one or more large ocellated spots in their dorsal fins.
● **Aquarium maintenance and breeding:** These fish are best treated like dwarf cichlids of the genus *Apistogramma*. Water should be soft and acid, and kept scrupulously clean and warm at 25-29°C (78-84°F). The addition of peat or peat extract to simulate blackwater is helpful. These are excellent community tank residents which prefer planted tanks.

Additionally, the inclusion of dither fish in the form of small schooling tetras will make them feel at home. Biparental substrate spawners whose spawns number 100-300. The fry can be somewhat small, so they should be provided with liquid fry food, microworms or rotifers before switching to newly-hatched *Artemia* nauplii.

An excellent choice for the small, planted community tank. Several colour varieties are available, including blue and red morphs. In addition, a second look-alike species, *L. dorsigera*, is available from the La Plata system (Rio Paraguay/Rio Parana), often in wild shipments direct from Argentina.

Although superficially resembling *L. curviceps*, this fish develops a bright maroon to fireengine red breast and face when brood tending. Care like *L. curviceps*, however, lower temperatures are possible given the subtropical origins of this fish. A third, as yet undescribed species, a mint-green curviceps-like fish from the Mato Grosso region of Brazil is also occasionally imported, as is the slightly larger form *L. flavilabrus* from Peru, and the 'giant curviceps', *L. thayeri*, from Brazil. Care for all 'Smiling Acaras' as for *L. curviceps*.

Above: *The 'smile' is easily seen on this* **Laetacara thayeri**, *the giant of the group.*

Below: **Laetacara curviceps** *is the most common of the 'Smiling Acaras'.*

MOUTHBROODING ACARAS
The Mouthbrooding Acaras, *Bujurquina* species, are a group of 17 species which practise delayed (primitive) mouthbrooding. Eggs are laid on a substrate, often movable, and are fanned and guarded for 24-48 hours. Then the larvae are chewed from their eggshells and uptaken into the mouths of one or both of the parents where they are safely incubated for a further 2-6 days.

The free-swimming fry commence foraging immediately and, if threatened, stream back into the open mouths of the parents. This intense parental care is practised for 6-8 weeks post-spawn.

These cichlids are elongate medium-sized at 10-15cm (4-6in), and often have a fringed caudal fin. In addition, they all have an obliquely-oriented black lateral band that runs from the eye back to the insertion of the soft dorsal fin, the latter visible even at small size. Many of the species are found in rather restricted areas which are not commerically collected. At least three species are sporadically available: *B. mariae, syspilus,* and *vittatus.* All are excellent for the planted community tank. The name *Bujurquina* (pronounced *boo-her k'eye na*) was created from the actual native name for these fish, *bujurqui.*

Above: **Bujurquina mariae** *is one of the more common of the Mouthbrooding Acaras.*

Bujurquina vittata

'Paraguayensis', Paraguay Mouthbrooder

● **Distribution:** Rio Parana/Rio Paraguay of the La Plata system (Paraguay, Argentina).
● **Length:** Medium-sized 10-13cm (4-5in) Acara.
● **Diet:** Frozen and live food supplements recommended.
● **Sexing:** Essentially isomorphic.
● **Aquarium maintenance and breeding:** Should be maintained as for the Smiling Acaras. These make good denizens of the medium-sized planted community tank as they are

Below: The 'Paraguayensis', **Bujurquina vittata**, *is a mouth-brooding Acara which originates from Paraguay and Argentina.*

somewhat shy. Can take lower temperatures. Spawns number 50-100 eggs which are uptaken by both parents. The free-swimming fry may need smaller first foods.

Originally known as *Aequidens paraguayensis* in the hobby, Kullander relagated that name to junior synonymy with *B. vittata*. However, when sporadically available, the fish is still sold under the name 'Paraguayensis'. In the wild, these and other *Bujurquina* species preferentially select waterlogged leaves as movable platforms for their spawns. When danger threatens or if water levels drop, they simply drag the spawn-bearing platform elsewhere.

Below: *The Keyhole Cichlid,* **Cleithracara maronii**, *is a good candidate for the planted South American community tank.*

Above: *This* **Krobia guianensis** *from the Guianas is usually sold to aquarists as 'Itanyi' or the Dolphin Acara.*

MISCELLANOUS ACARAS
The remaining Acaras have been installed in a variety of genera, several with only one species (monotypic). These include *Cleithracara, Guianacara, Krobia, Nannacara* and *Tahuantinsuyoa.* Several of these are common aquarium fish.

Cleithracara maronii

Keyhole Cichlid

● **Distribution:** The Guianas (Rio Maroni in Surinam).
● **Length:** Large males up to 13cm (5in), females a little smaller.
● **Diet:** Requires frozen or live food for optimal conditioning.

● **Sexing:** Essentially isomorphic.
● **Aquarium maintenance and breeding:** Best handled like *Laetacara curviceps* (above). Shy almost to a fault, these fish require planted tanks with plenty of shelter to feel comfortable. Dither fish, in the form of peaceful schooling tetras, recommended. A rich diet of frozen and live foods will bring them to ripeness. Moderately difficult to spawn; notorious egg eaters. Artificial incubation recommended for pairs which continually eat their spawns.

The genus *Cleithracara*, from *kleithron* (Greek for lock), commemorates the common name of this fish. Commercially bred in Asia, wild specimens are rarely imported.

Nannacara anomala

Golden-Eye Dwarf Cichlid
● **Distribution:** The Guianas, principally imported from Guyana.
● **Length:** Males up to 7.5cm (3in), females half that length.
● **Diet:** Undemanding omnivore.
● **Sexing:** Highly dimorphic. Males have high, rounded foreheads and iridescently spangled flanks. Females remain brown and express a parallel two-stripe flank marking.
● **Aquarium maintenance and breeding:** Truly a beginner's cichlid. Undemanding in terms of maintenance and diet: can be conditioned on prepared foods if necessary. Small size makes them a natural choice for planted community tanks. Several hundred eggs are laid on a receptacle and guarded assiduously by the female. When brood-tending, the parallel two-stripe pattern of the female is replaced by a distinctive latticework pattern: vertical bars intersecting the parallel stripes herald motherhood! The female is exceedingly intolerant of the male and will chase him off. In small tanks, the male may be killed and should be removed. The female takes exceptional care of her offspring who will grow even on crushed dry food. A great choice for anyone wishing to gain experience in breeding substrate-spawning cichlids.

Nannacara anomala is usually available at low cost as the principal species in boxes of 'Assorted Dwarf Cichlids' from Guyana. There are several colour geographical variants available, including orange, black and red.

A second species, *N. aureocephalis*, with a gold head and bigger body, has been described as from French Guiana, but it may well prove to be *N. anomala*.

Below: *The Golden-Eye Dwarf* **Nannacara anomala,** *(♂ pictured). An undemanding species and a wonderful beginner's cichlid.*

Left: **Guianacara sp. affin. geayi** *is an interesting cave-spawning cichlid.*

Guianacara sp. affin. geayi

'Geayi', Bandit Cichlid

● **Distribution:** Guianas, Venezuela.
● **Length:** Males 15-18cm (6-7in), females one-half to two-thirds as large.
● **Sexing:** Males develop a large, block-like nuchal crest.
● **Aquarium maintenance and breeding:** The requirements of *G. sp. affin. geayi* in the aquarium are few and they do well in the mixed, rowdy cichlid community tank. They are found in rocky pools in nature and are cave spawners in the aquarium. Thus they should be provided with some sort of cave, particularly a ceramic flowerpot with a drainage hole carefully enlarged to allow entrance. Eggs, often green, are laid on the vertical sides of the pot. Care of fry as usual. They will spawn precocially at a length of 75cm (3in).

'Geayi' (pronounced *gay'-eye*) has had a rather uncertain taxonomic history owing to its transitional status.

Originally installed in the genus *Aequidens, geayi* was moved first to *Acarichthys* (with *heckelii*) and most recently to its own newly-created genus *Guianacara*. Kullander, who did the moving, has described a series of species, in addition to *geayi*, which he has installed in the genus and which are distinguishable primarily on the basis of banding.

Apparently the true *G. geayi* is found only in French Guiana, thus is unlikely to be the hobby fish whose origin was Guyana. The common 'Geayi' is probably *G. sphenozona*, but the name *'geayi'* persists in the trade. A second, undescribed 'geayoid' from Venezuela, immediately recognizable from the striking bright orange-red spots on its cheeks (preoperculum), is also available in the hobby.

EARTHEATERS AND ALLIES

Few natural groupings of species provide the range of aquaristic challenge that the South American Geophagines or Eartheaters do. The genus *Geophagus* was 'created' by Heckel in 1840 for cichlids which shared a lobed first gill arch and small 'rakers' on the margin of their gills. These gill rakers allow for their habit of sifting the substrates for detritus. Mouthfuls of sand or mud taken by the elongate, underslung mouth are strained through the rakers, the food retained, and the inedibles passed out via the gill covers. Even the natives refer to them as 'Eartheaters'. The original genus *Geophagus* has recently been limited by Kullander, and the remaining Geophagine cichlids placed in other genera including *Gymnogeophagus, Satanoperca* and *Biotodoma*. Several species await reclassification into their own new genera, and these are designated as *'Geophagus'* species which are pending reassignment to new genera.

Within the assemblage there are three major modes of reproduction: simple substrate spawning; primitive (delayed) mouthbrooding, as in the Mouthbrooding Acaras; and immediate (advanced) mouthbrooding. In immediate mouthbrooding, pairing is brief and the female gathers the eggs, which are fertilized in her mouth, and incubates them to full term, much like her African counterparts from the Great Rift Lakes. Eartheater cichlids come in all shapes, sizes and dispositions.

Satanoperca leucosticta

Demonfish, 'Jurupari'

● **Distribution:** Regions of Guyana, Amazonia.

● **Length:** Up to 30cm (12in).

● **Diet:** In nature, these fish sift benthic invertebrates and plant material ('grut') from the substrate. In the aquarium they are omnivorous, but require frozen and live foods for optimal conditioning. They prefer sifting over fine gravel or sand. Sinking or pelleted foods recommended.

● **Sexing:** Essentially isomorphic. In the wild, the male of the pair can usually be identified as the larger fish (Cichocki 1976, Lowe-McConnell 1969).

● **Aquarium maintenance and breeding:** Like most *Satanoperca* species, *S. leucosticta* demands clean, warm water, around 25-29°C (78-85°F). Demonfish prefer soft, acid water and benefit from the addition of peat filtration or peat extracts. A tangle of bogwood and non-anchored or floating plants together with dither fish will make them feel less shy. While other cichlids may be housed with them, these should be peaceful: *S.leucosticta* may be inhibited from spawning in the presence of other cichlids.

Satanoperca leucosticta is a biparental, delayed mouthbrooder. It is moderately difficult to induce to spawn in the aquarium. In the wild, pairs have been observed choosing movable platforms such as waterlogged wood or even sneakers, which they tug around with them in response to potential predation. In the aquarium, they will accept stones. About 150-300 eggs are laid and then covered with a thin layer of gravel or sand. The eggs are fanned and guarded for about 48 hours at which point the parents chew the larvae out of their eggshells and uptake them for further incubation in their mouths. Both parents participate in this buccal incubation, taking turns and passing the larvae back and forth for several days. When free-swimming, the fry are released to swarm and forage, but the parents continue to provide buccal shelter for several weeks thereafter. When alarmed, one or both parents will assume a head-down, mouth-extended posture and the fry respond by swarming to and diving into the open mouth for protection. When danger is past, the fry will be spat out or blown backwards through the gill covers (opercula) and resume their foraging. After about 3-4 weeks, the fry and parents should be separated. The free-swimming fry eagerly eat newly-hatched *Artemia* nauplii as their first meal, but grow somewhat slowly.

Known in the aquarists' hobby as *Geophagus jurupari*, the silvery,

Below: **Satanoperca leucosticta,** *the 'Jurupari' of the hobby.*

spot-faced 'Jurupri' was more correctly identified as *Satanoperca leucosticta* by Kullander in 1986. The true *S. jurupari* is a plain silver or gold eartheater with an unspotted face is is not commonly available in the trade. There is at least one other unspotted 'juruparoid' eartheater – *S. pappaterra* – from the Brazilian Mato Grosso than can be distinguished from these other two by virtue of its gold base coloration and distinctive longitudinal black/brown stripe that extends from the eye to the tail. All are maintained as above.

The name 'Jurupari', applied both commonly and scientifically to these fish, is a native Tupi name for a feared forest demon. The natives called these fish 'juruparipindi', or 'demon's lure' (fishhook), but the relationship of this fish to the Jurupari myth is obscure. Kullander (1986) commemorated this relationship by resurrecting the genus *Satanoperca* Guenther, (1862), which translates from the Latin as 'Satan's Perch'. Hence the common name 'demonfish' for this group of interesting cichlids.

Satanoperca daemon

Daemon, Three-Spot Demonfish

● **Distribution:** Colombia; Venezuela; Rio Negro, Brazil.
● **Length:** Can grow up to 30cm (12in).
● **Diet:** Sifting detritivore, handled like *S. leucosticta.*
● **Sexing:** Essentially isomorphic. Both sexes with multiply-filamentous dorsal fins.

Above: *The Three-Spot Demonfish,* **Satanoperca daemon,** *sports a diagnostic caudal ocellus.*

● **Aquarium maintenance and breeding:** Care is identical to that of *S. leucosticta* with a little more attention to water chemistry. *S. daemon* is notoriously difficult to rear, and has been successfully propagated in captivity only a few times. These fish are susceptible to 'Neotropical Bloat', a terminal condition that involves extreme blockage and bloating of the belly region. The origin of the condition is unclear; however, strict attention to water quality is essential. Dissolved oxygen may be an important factor and wet/dry trickle filters are recommended. Spawning has been achieved in simulated blackwater having a pH of 4.5 and no measurable hardness; the use of RO-processed water is recommended. Although all *Satanoperca* species are believed to be biparental, delayed mouthbrooders, *S. daemon* may not be. In one of the very few spawning accounts published, Eckinger (1987) reports that the species displayed modified substrate spawning.

S. daemon is easily recognizable by the white-ringed black ocellus on the caudal peduncle, and the two mid-lateral blotches, one in the centre of the fish just below the lateral line, the second equidistant between it and the ocellus. However, there are two other 'Spotted-Juruparoids' that offer some confusion. One of these is the newly-described *S. lilith* that enters the hobby as a rare contaminant from the Brazilian Amazon. This fish resembles *S. daemon* in all ways except that it has a single mid-lateral blotch located on and just above the lateral line. A more-commonly encountered, but nevertheless rare, lookalike from Brazil is *S. acuticeps. S. acuticeps* is distinctive for the three equally-spaced black mid-laterial blotches that grace its flanks and the presence of a simple black blotch, rather than an ocellus, on the caudal peduncle. All three species develop spectaculor long red, multiple (five) filaments on their dorsal fins.

Neither *S. lilith* nor *S. acuticeps* have been spawned in captivity, nor are they regularly available to the hobby; however, they are sometimes mixed in with *S. leucosticta* in wild shipments from Brazil. Care and approach for all should be like that of *S. daemon*.

Geophagus proximus

'Flag Tail' Surinamensis

- **Distribution:** Amazon basin.
- **Length:** Up to 30cm (12in).
- **Diet:** Omnivorous.
- **Sexing:** Essentially isomorphic.

Below: **Geophagus proximus** *is one of the 'Surinamensoid' Eartheaters.*

- **Aquarium maintenance and breeding:** In general, the parameters established for *S. leucosticta* apply equally well here. *Geophagus proximus*, and other related 'surinamensoids', are somewhat more aggressive cichlids which can be housed and bred in a community situation. They are best kept as a group of several individuals to diffuse the aggression. Although water quality must be maintained, water chemistry seems less important than for the Satanopercoids.

G. proximus is an advanced mouthbrooder with the female uptaking the eggs. The male may or may not participate in the rearing and defence of the fry. Newly-hatched *Artemia* nauplii are a fine first food, and growth is rapid. They are reproductively competent at a size of 11.5cm (4½in), at an age of 1-1½ years.

This fish, and several related species, have been known in the trade for years as *Geophagus surinamensis* – a fish initially believed to be cosmopolitanly-distributed with a number of geographic colorational morphs. However, aquarium-based observations of spawning behaviour have suggested that several discrete species make up the 'surinamensoid complex'. While the black-chinned Guyanese form,

Above: *This male Red Hump Eartheater,* **'Geophagus' steindachneri**, *displays the characteristic conspicuous red nuchal hump.*

subsequently named *G. brachybranchus* by Kullander, is a delayed mouthbrooder, several other Amazonian forms (*G. altifrons megasema*), including *G. proximus*, have proven to be immediate mouthbrooders and one, the recently described *G. argyostictus* from the Rio Tocantins, is even a simple non-mouthbrooding substrate spawner. Although readily distinguishable on the basis of colour pattern, *G. proximus* and most of the other surinamensoids, are still sold as *G. surinamensis* in the trade. The true *G. surinamensis*, from Surinam, has yet to be imported commercially.

'Geophagus' steindachneri

Red Hump Eartheater

● **Distribution:** Colombia, Venezuela.

● **Length:** Males up to 15cm (6in), females half to two-thirds of this size.

● **Diet:** Omnivorous.

● **Sexing:** This fish is named after one aspect of its well-defined sexual dimorphism: the large, red nuchal hump that mature, dominant males sport on top of their heads. Females lack any hint of the hump. In addition, females retain half-two-thirds the size of their consorts, and are usually drably coloured. Males typically have some modest iridescent spangling (green, black, orange) on their flanks. The species is known to be sexually precocious at small size.

● **Aquarium maintenance and breeding:** Although Red Humps, particulary dominant males, can be aggressive, these are excellent beginner's cichlids. They are totally undemanding in dietary and water requirements and will spawn at small size (females 2.5-4cm (1-1½in). These are immediate, maternal mouthbrooders. Courtship is brief and pair-bonding non-existent. Ripe females select and clean off suitable substrates to hold their spawns and then solicit the male to participate. Eggs are laid a few at a time with the female backing up to scoop them up in her mouth. The male spreads his milt over the substrate and, presumably, sperm is picked up by the female when she scoops up the next batch of eggs. As many as 150 eggs may be laid, dependent on the size of the female, but spawns typically number 30-50. Ovigerous females are excellent single parents and rarely lose a spawn. They should be moved to separate quarters for release of the free-swimming fry 8-10 days post-spawning. As they do not eat during incubation, the females should be kept with their brood and fed for several days before being returned to the colony. Males are harem polgynists capable of spawning with several females in succession, which are best maintained as a large harem. Females kept as a harem

will ripen together and spawn nearly synchronously. Fry of this species are large in size and easy to raise, even when fed on crushed dry or prepared food.

Regrettably, this fish is availiable in the hobby under an assortment of incorrect names, including *'G.' pellegrini* and *'G.' hondae*. *'G.' pellegrini* is the valid name of a larger, 'Humped' Eartheater that hails from the Pacific slope of southwestern Colombia up to Panama, and which has rarely been imported into the hobby. *'G.' hondae* is a junior synonym of *'G.' steindachneri*, and is therefore incorrectly used. A third humped species, *'G.' crassilabrus*, is found in Panama, and has entered the hobby only through the personal efforts of dedicated amateur aquarists who have collected it there. Like *'G.' pellegrini*, it is a

Below: *The Mother-of-Pearl Eartheater* **'Geophagus brasliensis'** *is an easy, beginner's cichlid.*

Above: *The Thick-Lipped Eartheater* **'Geophagus' crassilabrus**, *is a rare 'Humped' Eartheater from Panama.*

larger species reaching 25cm (10in) in length. All three 'Humped' Eartheaters are sexually dimorphic harem polgynists, and all three are immediate mouthbrooders. Since *'G.' steindachneri* occurs in several colour varieties (black, orange, green) and since neither *'G.' pellegrini* nor *'G.' crassilabrus* are regularly imported, the 'Red Hump' found in the trade is most likely to be *'G.' steindachneri.*

'Geophagus' brasiliensis

Mother-of-Pearl Eartheater

● **Distribution:** Southern Brazil, Argentina.
● **Length:** Males up to 25cm (10in), females half to two-thirds size.
● **Diet:** Omnivorous.
● **Sexing:** Mature males develop pronounced nuchal humps, females do not. This particular species is known to be sexually precocious and sexable at small size.

● **Aquarium maintenance and breeding:** *'Geophagus' brasiliensis* is a beautiful cichlid with modest requirements. Hailing from subtropical South America, these fish can withstand temperatures of 10°C (50°F) while tolerating more tropical aquarium conditions. Their coastal distribution suggests that harder, alkaline water is to their liking and the addition of crushed dolomite or oyster shell to their filter box will aid in increasing the carbonate hardness and buffering the pH. They can be conditioned on a diet of good prepared foods, although this is not advised. They are typical substrate spawners laying several hundred eggs and are exemplary parents. Fry are easily raised on a diet of crushed dry food and newly-hatched *Artemia*. Growth is rapid. A good beginner's first experience in breeding a typical substrate spawning cichlid.

The Mother-of-Pearl cichlid is a truly beautiful fish. Each scale centre is marked with a blue, green or silver nacreous spot on a base colour that ranges from brown to mahogany to bright red, while the unpaired fins are spotted and striped with hyaline dots. Mature males, which can easily reach 25cm (10in) in captivity, are truly breathtaking. Unfortunately, juvenile *'G.' brasiliensis* are brown and fail to develop adult coloration until they grow to nearly 7.5-10cm (3-4in). This, coupled with their ease of propagation and the size of the spawns, has spelled commercial doom for this fish. Although inexpensive, they are well worth the space and attention in even the advanced cichlid hobbyist's collection for their sheer beauty.

G.' brasiliensis is highly polymorphic. Many of the differently coloured populations may well prove to be valid species in their own right. One fish that has been confused, historically, with *'G.' brasiliensis* is *Gymnogeophagus gymnogenys*. Although earlier pictures suggested an elongated, 'brasiliensis'-like fish, more recent collections in southeastern Brazil have yielded a pearl-scaled eartheater that is a delayed mouthbrooder, the 'real' *G. gymnogenys*. This latter fish is available sporadically in the hobby and is of particular interest to cichlid aficionados for its beauty and rarity.

Below: **Gymnogeophagus gymnogenys**, *a delayed mouthbrooder, has often been confused with* **G.' brasiliensis'.**

Gymnogeophagus balzanii

Paraquay Eartheater, 'Balzanii'

● **Distribution:** La Plata Drainage, Argentina, Paraquay.

● **Length:** Males up to 20cm (8in), females half to two-thirds of this size.

● **Diet:** A snail crusher in the wild. Omnivorous in the aquarium.

● **Sexing:** Dramatically dimorphic. Dominant, sexually mature males, as small as 5cm (2in), develop huge nuchal hoods on their forehead giving the fish a 'blockheaded' look. Males also express 4-5 parallel metallic blue longitudinal stripes on their flanks and have large, fan-like ventral fins spotted in blue. Females lack the spangling, have smaller yellow ventral fins, remain half to two-thirds the size of males and never develop the hood. The size differential between the sexes is apparent early on.

● **Aquarium maintenance and breeding:** *Gymnogeophagus balzanii* is the canary of the neotropical cichlid world: water must be kept scrupulously clean. When water quality declines they are particularly susceptible to neuromast erosion or pitting, known as 'Head Hole'. Hailing from subtropical regions of Paraquay and Argentina, they can take temperatures down to 15-17°C (low 60s°F), but are best maintained at 24-27°C (76-80°F). Regular prepared, frozen and live foods keep these eager eaters in the best of health. They can be aggressive with each other and with other cichlids.

G. balzanii is a harem polgynist, best kept and maintained as a colony of one male and two or more females.

Above: *Male* **Gymnogeophagus balzanii** *is notable for the huge nuchal hump it develops when sexually mature.*

Ripe females establish and defend territories with a spawn receptacle as the focal point. At 24 hours prior to egg-laying, they assume brood care coloration which consists of a dark mid-lateral blotch, bandid eye-cheek stripe, and blackening of the edges of the ventral fins. The male's participation in the process is brief: these fish are known to be delayed, maternal mouthbrooders.

The free-swimming fry are too small to take newly-hatched *Artemia* and must be offered liquid fry foods, microworms or rotifers for the first few days. Growth is rather slow. Sexual maturity is reached at a size of 5-7.5cm (2-3in) which is attained in about one year.

The genus *Gymnogeophagus* was established by Ribeiro in 1918 to accommodate eartheater species lacking cheek scalation. Goose, in 1975, revitalized the genus and described several further characters which distinguish it from the other geophagines. There are currently eight species in this genus of 'naked eartheaters', including both substrate spawners and delayed mouthbrooders.

Biotodoma cupido

Cupid Cichlid

● **Distribution:** Region of the Amazon drainage.
● **Length:** Up to nearly 15cm (6in).
● **Diet:** Requires frozen and live foods.
● **Sexing:** Reportedly dimorphic with respect to the iridescent blue, vermiform markings that develop on the face of mature specimens. In males, these are lines, whereas females have spots. Otherwise identical in size and finnage.
● **Aquarium maintenance and breeding:** A somewhat delicate fish best treated like the *Satanoperca* species. Keep them warm at a temperature of 27-29°C (80-84°F), clean, and in soft, acid water (pH 5-6, less than 1°dH) – fine in a planted tank. They tend to be somewhat scrappy so shelter in the form of bogwood or other is recommended. Feeding can be a problem initially but newly-hatched *Artemia* nauplii are eagerly consumed. Once eating, frozen or live bloodworms, glassworms and/or mosquito larvae are taken readily. Spawning accounts are few. Ripe pairs excavate a pit in the gravel and lay about 100 eggs on the bottom. The free-swimming fry are too small for newly-hatched *Artemia* nauplii and growth is slow. The use of RO-processed water will increase the probability of spawning.

The genus *Biotodoma* was separated from the other Geophagines on the basis of their smaller snouts and the positioning of the mouth. The genus contains one other described, and several undescribed, species. *Biotodoma wavrini*, from Guyana and the Orinoco basin, may be distinguished from *B. cupido* by the positioning of the flank ocellus: on and above the upper lateral line just below the dorsal fin in *B. cupido*, below the lateral line near the body's midline in *B. wavrini. B. wavrini* is also slightly more elongated. Care as for *B. cupido.* Usually imported at small size, young *Biotodoma* are typically hollowed-out, grey, nondescript fish. However, with proper care and feeding, they metamorphose into spectacularly contoured adults at about two years of age.

Below: **Biotodoma cupido** *is a beautiful but demanding cichlid that has rarely been spawned.*

Acarichthys heckelii

Heckel's Threadfinned Acara

● **Distribution:** The Guianas, and the region of the Amazon drainage.
● **Length:** Up to 20cm (8in).
● **Diet:** Omnivorous.
● **Sexing:** Essentially isomorphic.
● **Aquarium maintenance and breeding:** Relatively undemanding, *A. heckelii* tolerates a variety of water chemistries and eats anything. They can be belligerent, so careful choice of tankmates and adequate hiding places are essential.

In the wild, *A. heckelii* has a most interesting spawning behaviour. Females excavate a series of tunnels in the soft mud of the bottom that lead to a larger 'nuptial chamber', a hollowed-out cave where the eggs will be laid. Once finished, the females actively court males swimming into their territories. Successful pairing results in the attachment of nearly one thousand eggs to the walls and ceilings of the nuptial chamber. These are fanned by the female while the male provides perimeter defence topside. Once hatched and free-swimming, the tunnel system provides the focal point for their foraging and, if threatened, parents and fry retreat into its safety. This unusual spawning mode may be approximated in the aquarium using a large inverted clay flowerpot or large-diameter PVC pipe stood on end. Although difficult to spawn, high temperatures of 29-32°C (85-90°F) and frequent, large water changes seem to initiate spawning. Fry are easy to raise.

Above: *Heckel's Threadfinned Acara,* **Acarichthys hecklii** *excavates tunnels as part of its unusual spawning behaviour.*

Although closely allied to the Eartheaters, *A. hecklii* lacks a lobed gill arch. The species *geayi*, formerly in *Aequidens*, was placed in *Acarichthys* for a while. More recently, Kullander (1989) has created the genus *Guianacara* for *geayi* and several *geayi*-like fish from the Guianas. *A. hecklii*, with its dramatically produced 'threadfinned' dorsal, is one of the most spectacular neotropical cichlids in the hobby. Look for its as a contaminant in shipments of *S. sp.* 'jurupari' and/or *G. sp.* 'surinamensis'.

APISTOGRAMMA AND ALLIES
Members of the genus *Apistogramma*, or Apistos for short, are dwarf cichlids ranging in size from 2.5-7.5cm (1-3in). Like Eartheaters, *Apistogramma* species have a lobed first gill arch with rakers arranged along the inside margin. However, the positioning of their lateral line is closer to the dorsal fin than that in the Geophagines. Despite their affinity with the Eartheaters, Apistos do not sift; rather the pick at small benthic invertebrates, chiefly insect larvae and worms, which form the main part of their diet in the wild. In the aquarium, they require live or frozen foods for optimal health. There are nearly 50 described species of *Apistogramma*, and a handful of undescribed forms found all over tropical South America. They live in streams, ponds and oxbow lakes off the main rivers over fine bottoms, often with submerged branches and considerable leaf litter which are used for shelter and the caves in which they spawn. Care for the group is virtually identical and is described below for three available Apistos.

In addition to the Apistos, there are a variety of other dwarf cichlids which are closely allied. These include members of the genera *Microgeophagus (Papiliochromis), Apistogrammoides, Taeniacara, Biotoecus, Crenicara, Dicrossus* and *Mazarunia.*

Several of these, particularly the Rams (*Microgeophagus* species) and the Chequerboards (*Crenicara Dicrossus* species), are popular, beautiful aquarium fish the maintenance and feeding requirements of which are virtually identical to the Apistos.

Apistogramma agassizii

Agassiz's Apisto

● **Distribution:** Blackwaters of the Peruvian and Brazilian Amazon.
● **Length:** Males up to 7.5cm (3in), females 2.3-3.75cm (1-1.5in).
● **Diet:** Requires frozen and live foods.
● **Sexing:** Highly dimorphic. Males nearly twice as large as females, with more highly developed finnage. Their tails are spade-like, and their colours more vibrant. Spawning/brooding females turn bright golden with conspicuous black trim to their ventral fins, a black 'bandit' eyeband, and darkened lateral spot.
● **Aquarium maintenance and breeding:** Like most Apistos, these are best handled in planted tanks with peaceful schooling dither fish like tetras, pencilfish, etc.

Should be kept clean and warm in temperatures of 26-29°C (78-85°F). Sponge filters are a help in maintaining water quality, as are regular water changes. Best success with most Apistos is by using soft, acid (pH 5-6) water; the use of RO-processed water can be helpful. The addition of peat extract or boiled peat in the filter is recommended to simulate the conditions of blackwater.

Food should be primarily frozen and live, with bloodworms, mosquito larvae, and glassworm larvae alternating. A 'cave' for spawning (an inverted flowerpot with a notched rim, hollowed-out half coconut shell, rock pile) should be provided. In larger tanks of 90-180 litres (20-40 gallons), multiple pairs, or one male and several females, may coexist peacefully with one another. Each female in the harem should be provided with her own cave.

Below: *Male* **Apistogramma agassizii** *are known by their conspicuous spade-shaped tails.*

● **Comments:** Named after the eminent natural scientist, Louis Agassiz of Harvard, the name is correctly pronounced *'Ag-a-see'-eye'*. Several colour variants (local populations) are available in the hobby, including red, gold and blue. These colours are predominently displayed in the dorsal, anal and caudal fins of the males.

Apistogramma cacatuoides

Cockatoo Apisto

● **Distribution:** Streams and oxbow lakes of the Peruvian Amazon.
● **Length:** Males to 9cm (3½in), females half that size.
● **Diet:** Requires frozen and live foods.
● **Sexing:** Highly dimorphic. Male develops magnificent 'cockatoo crest' in which the first 3-5 spiny rays of the dorsal fin are elongated dramatically. The tail fin of the male is moderately lyrate and may have one or several irregularly coloured spots. Females lack these particular features.
● **Aquarium maintenance and breeding:** Maintenance as for *A.agassazii*. Cave-spawning, harem polgynists best maintained in groups of one male to several females. Each female should be provided with her own cave and territorial space. Harems should be maintained in tank of 140 litres (30 gallons) or larger, but may be spawned in pairs in smaller tanks, with care taken to provide shelter and dither or target fish.

Like *A. agassizii*, the Cockatoo Apisto is available in several colour morphs which reflect its patchy distribution in the wild. One of these, a red form, is particularly notable for the extensive red spotting that develops in the caudal fin of the male. These are being captively bred and selected for increased colour and spotting in Germany. Wild *A. cacatuoides* are available occasionally in shipments of mixed dwarf cichlids from Peru.

Apistogramma steindachneri

Steindachner's Apisto

● **Distribution:** The Guianas.
● **Length:** The largest of the hobby Apistos; males can reach 11cm (4½in), females 2.5-5cm (1-2in).
● **Diet:** Omnivorous.
● **Sexing:** Males considerably larger than females with slightly lyrate, filamentous tipped tail.
● **Aquarium maintenance and breeding:** The most common and easily maintained of the Apistos. While the stringent conditions outlined for *A. agassizii* benefit this fish, Steinachner's Apisto can be maintained and bred on prepared and frozen foods with a little less concern for water chemistry and condition. May be bred in pairs, but may also be polgynous if provided with multiple females.

A high-bodied, robust Apisto typically

Below: *Males of the Cockatoo Dwarf,* **Apistogramma cacatuoides** *develop dorsal fin cockatoo crests.*

Above: *Steindachner's Apisto,* **Apistogramma steindachneri**, *is among the easiest of the dwarf cichlids to maintain in the aquarium.*

mixed with *Nannacara anomala* in mixed dwarf cichlid shipments from Guyana. They are readily available, inexpensive, hardy and provide a good beginning point for learning about Apisto feeding, care and breeding.

Named after the early twentieth-century Austrian ichthyologist and student of cichlids, Franz Steindachner.

Microgeophagus ramirezi

Ram, Ramirez's Dwarf, Butterfly Dwarf

● **Distribution:** Streams and ponds of the Rio-Orinoco drainage in the Venezuelan and Colombian *llanos.*

● **Length:** Can grow up to 6cm (2½in).

● **Diet:** Frozen and live food recommended.

● **Sexing:** Moderately dimorphic. 'Cockatoo crest' (elongate first several rays of the spiny dorsal fin) of the male slightly more produced than that of the female. Females slightly smaller and rounder, and when ripe develop a rosy red flush to their ventrum.

● **Aquarium maintenance and breeding:** Best maintained as for *A. Agassizii.* Unlike most Apistos,*M. ramirezi* is monogamous and spawns in the open, typically on a stone. Fry are often difficult to raise with parents and may require artificial incubation. They are small

Below: *The Venezuelan Ram,* **Microgeophagus ramirezi,** *is a beautiful dwarf cichlid.*

and may require liquid fry food, microworms or rotifers for their first feedings. Often high temperatures of 30-31°C (86-88°F) are necessary to elicit spawning.

First captured and bred by Manuel Ramirez, for whom the fish was named, and Herman Blass in 1947. Initially described in the genus *Apistogramma*, but subsequently removed to the genus *Microgeophagus*. Kullander later proposed the genus name *Papiliochromis* for the Ram, but his position is controversial and, for the time being, *Microgeophagus* would appear to be the correct generic nomen. Aquarists will find additional information in the hobby literature under all three of these names.

A second species, *M. altispinosa*, the Bolivian Ram has recently appeared. A larger, more Geophagine-like fish with a green/gold cast to the body, *M. altispinosa* grows to 10cm (4in) and, like its congener, is an open substrate (non-cave) spawner.

Both species are regularly bred in aquaria in Asia, where selective breeding has produced several cultivars of the Ram including golden (xanthistic) and 'veiltail' varieties.

Dicrossus filamentosus

Lyretailed Chequerboard

● **Distribution:** Blackwaters of the Rio Negro, Brazil, and the Rio-Orinoco basin of Colombia.
● **Length:** Males up 7.5cm (3in), females two-thirds that size.
● **Diet:** Frozen and live foods recommended.
● **Sexing:** Dramatically dimorphic. Males develop lyrate tail fins produced to filamentous streamers on top and bottom and iridescent blue spangled stripes over the central chequerboard band. Females lack the tail streamers and iridescence.
● **Aquarium maintenance and breeding:** Care as for *A. agassazii.* The Lyretailed Chequerboard has been bred infrequently and with considerable difficulty. Should be kept warm, 30-31°C (86-88°F), and in very clean water. As they live in blackwaters with cardinal tetras (*Paracheirodon axelrodi*), the water should be very acid (pH 4-5) and have little, if any, hardness: RO-processed water is helpful. Additionally, peat extract helps to

Below: *The Lyretailed Chequerboard,* **Dicrossus filamentosus,** *is among the more challenging and beautiful of the dwarf cichlids.*

Above: **Crenicara punctulatum** *is a larger, higher-bodied chequerboard cichlid.*

duplicate in blackwater environment. The fish often select broad-leafed plants, which are recommended, to support their clutches of 50-150 eggs. Brooding females are known by a particular dark black longitudinal stripe down the middle of their body.

Eggs often fail to hatch, probably due to inappropriate water chemistry. Both fry and adults benefit from

feedings of newly-hatched *Artemia* which is an excellent 'first food' for newly-imported, emaciated specimens.

Described at first as *Crenicara filamentosa* by Ladiges in 1958 from aquarium specimens, Kullander re-described the species in 1978, and has since advised placement in the separate genus *Dicrossus*. Most aquarium references use the older name. There are two other 'chequerboard cichlids' that aquarists may encounter. *Crenicara punctulatum* is a much larger, more robust relative from Peru that may grow to nearly 15cm (6in). They share the distinctive chequerboard pattern, but they are high-bodied with ovate tails. *Dicrossus maculatus* is a transitional form that hails from the Brazilian Amazon. It shares with *C. punctulatum* the larger size of 10-13cm (4-5in) and an asymmetrical ovate tail, but is not nearly as high bodied.

Juvenile *D. maculatus* are virtually indistinguishable from *D. filamentosus*. Although the only chequerboard dwarf in the hobby of the 1940s and 1950s, *D. maculatus* has been absent until only recently (1988) when its location was rediscovered by German aquarists and breeding stock returned there. Care of all three species is identical.

PIKE CICHLIDS

Pike cichlids, of the genus *Crenicichla* are elongate, torpedo-shaped fish admirably adapted for life as piscivorous ambush predators. They have long snouts with large, teeth-studded protrusible mouths that enable them to grab and swallow smaller fish they vacuum in. Their common name is a tribute to the ultimate ambush fish, the pike (*Esox* species), however, they are related only in the convergence of their anatomy and habits. Pike cichlids are distributed throughout South America and number at least 50 species, including some dwarf forms. Despite their reputation as nasty aquarium residents, in fact the majority of pike cichlids can be kept in the rowdy cichlid community tank with other fish large enough to escape ingestion. Bonded pairs are particularly loyal and exceedingly gentle parents to their offspring, caring from them for 6-12 months in the wild. The tips on maintenance and breeding of the four species reviewed below hold for most pike cichlids.

Crenicichla sp. affin. saxatilis

Spangled Pike

● **Distribution:** Ponds, oxbow lakes, and swamps of the Guianas.
● **Length:** Males up to 30cm (12in), females about two-thirds as long.
● **Diet:** Initially live feeder fish, later freeze-dried krill and frozen foods.
● **Sexing:** Dimorphic. Males have extensive gold or silver spangling on their flanks and spotting in their dorsal and anal fins, both of which are produced to long filaments. Females have less spangling, have a distinctive white submarginal band in the dorsal fin sometimes with multiple, small ocelli, and develop distended cherry-red bellies when ripe.
● **Aquarium maintenance and breeding:** Pike cichlids, in general, are somewhat undemanding in the aquarium. Although predatory in nature, with a little patience they can be converted to a diet of freeze-dried krill, frozen bloodworms, and even some pelleted foods.

For emaciated, newly-imported specimens, live feeder fish will restore health until dietary conversion is attempted. Adults may be belligerent, so potential spawning partners should be separated until the female ripens and the divided 'pair' show interest in each other. Juveniles may be raised as a group if they are of the same size; smaller individuals will be picked on. Provide sufficient shelter in the form of PVC piping; at least one tube per fish.

Pike cichlids are cave spawners. In the aquarium acceptable caves include inverted notched flowerpots or stack driftwood. Several

Below: *This male Spangled Pike,* **Crenicichla sp. affin. saxatilis,** *has more body spangling than his female consort.*

hundred eggs are attached to the cave via adhesive threads, much like the eggs of West African cichlids (*Pelvicachromis*, *Nanochromis*) species. Free-swimming fry are huge, growth is rapid and the youngsters are soon eating diced frozen bloodworms and each other. The parents are unusually protective and gentle, and may continue to provide care for many weeks post-spawning.

The *saxatilis*-complex contains about 13 or so species which share similar coloration. *Crenicichla saxatilis* is limited in its distribution to the commercially-uncollected areas of Surinam and French Guiana, so its availability in the hobby is doubtful. However, *Cr. albopunctata* from Guyana is nearly identical in appearance and is the 'saxatilis' of the hobby. Several *saxatilis*-complex species, like *Cr. proteus* and *Cr. anthurus*, hail from the Peruvian Amazon, and at least one, *Cr. geayi*, is imported from the Rio-Orinoco. All have similar requirements in the aquarium.

Crenicichla regani

Dwarf Pike

● **Distribution:** Brazilian Amazon, Blackwater, Rio Tocantins.
● **Length:** Males can grow up to 15cm (6in), females up to 10-12.5cm (4-5in).
● **Diet:** Small feeder fish, freeze-dried krill and frozen foods.
● **Sexing:** Females alone sport 1-3 large irregular black/white occelli-like splotches in their dorsal fin, and have distended cherry-red

Above: *This female Dwarf Pike,* **Crenicichla regani**, *is conspicuous for the irregular black and white dorsal fin splotching.*

bellies when ripe. Females about two-thirds the size of males.
● **Aquarium maintenance and breeding:** Dwarf Pike cichlids are for aquarists with smaller, planted tanks and may be housed with a variety of characin dithers, as long as these are large enough to escape predation. These should be treated like Apistos and kept in soft acid blackwater, which is kept warm 26-29°C (80-85°F) and very clean. RO-processed water may be helpful in inducing spawning. They readily accept frozen bloodworms and prosper on them as a staple of their diet. Live dwarf red earthworms may be useful for conditioning.

These are cave spawners which have been spawned infrequently. Ripe females develop the distended, cherry-red bellies that signal their intent. Courtship involves head-down, belly-wriggling dancing reminiscent of courtship in the West African Krib, *Pelvicachromis pulcher*.

There are several species of Dwarf Pikes that grow less than 15cm (6in). These include the Amazonian species *Cr. notophthalmus*, whose females replace the irregular dorsal splotching of *Cr. regani* with several white-ringed-black ocelli in their dorsal fins, and two from Guyana: *Cr. wallacii* and *Cr. nanus,* which have not yet been in the hobby. Neither of the latter two species is reported to have dorsal fin ocelli. *Crenicichla heckelii* is the smallest of the dwarfs, reaching only 6cm (2½in), but this fish has not been imported from the Rio Trombetas into the hobby, nor has *Cr. urosema* from the Rio Tapajos rapids. The beautiful Dwarf Pike, *Cr. compressiceps*, from the Rio Tocantins/Xingu system, has appeared only recently. It grows to only 10cm (4in) and has been spawned in the aquarium. Dwarf Pike cichlids are definite 'must haves' for neotropical cichlid enthusiasts.

Crenicichla sp.

Orange Pike, French Fry Pike, Xingu Pike.

● **Distribution:** Rio Tocantins/ Xingu system.

● **Length:** Member of the *'strigata'*-complex, which reaches lengths approaching the region of 46cm (18in).

● **Diet:** Feeder fish, freeze-dried krill, pelleted prepared foods.

● **Sexing:** Females develop distended cherry-red bellies when ripe, and retain white submarginal bands in the dorsal and caudal fins. There is no apparent sexual difference in size.

● **Aquarium maintenance and breeding:** Despite their huge adult size, most members of the small-scaled *'strigata'*-complex are peaceful denizens of the large-fish community tank. Bonded pairs are devoted to one another and remain quite peaceful. Spawning has been achieved only infrequently, in a limited number of species. Those spawned, like *Cr. marmorata* have also proved to be cave spawners. Care as for *Crenicichla sp. affin. saxatilis.* These are easily converted to krill or pelleted foods. Members of the *'strigata'*-complex are best acquired in groups of 4-6 juveniles, of near identical size, and raised to adulthood. Be sure to provide enough shelter in the form of PVC tubing.

Species of the *'strigata'*-complex are notable for their shared juvenile colorational pattern, and for the dramatic colorational metamorphosis they undergo when maturing at around 15cm (6in) in length. Species of the complex include *Cr. cincta, funebris, johanna, lenticulata, lugubris, marmorata, ornata* and *strigata*, but are typically sold, particularly the young, as 'johanna', 'lugubris' or 'strigata'. Part of the excitement of raising these fish is seeing just into what the juveniles transform at 15cm (6in)! A good choice for breeders looking for a challenge, the Orange Pike has yet to be captively spawned.

Below: *This as yet undescribed* **Crenicichla** *species from the Rio Xingu is often sold as 'Orange Pike' when small.*

Above: *All juvenile pikes of the 'strigata' complex, like this 'Orange Pike', express a shared striped pattern until metamorphosing at 15cm (6in).*

Crenicichla sedentaria

'Frog-eyed' Pike, 'Hopping' Pike, Sedentary Pike

● **Distribution:** Peruvian Amazon, Ecuador, Colombia.
● **Length:** Males to 25cm (10in). Females half to two-thirds of this size.
● **Diet:** Small feeder fish, freeze-dried krill, pelleted foods.
● **Sexing:** Conspicuously dimorphic. Females with large ocellus in the dorsal fin. Ripe females develop a distended belly and bright red dorsal fin.
● **Aquarium maintenance and breeding:** As for the generality of pike cichlids, the 'Frog-eyed' pikes are believed to be cave spawners. PVC piping and/or inverted and notched flowerpot saucers, large enough to permit access and room to spawn, are recommended. These cichlids have not yet been spawned in the aquarium.

The genus *Batrachops* was created to hold a subgroup of pike cichlids with a unique arrangement of teeth and thick 'salami'-like bodies. The short snout confers a bug-eyed frog-like appearance to these fish, hence the genus name (batrachians are amphipians). Recently, Kullander has restored species of the genus *Batrachops* to the genus *Crenicichla*, but the former occupants of that genus form a coherent subgrouping of species. The species include: *Cr. cyanotus* (Amazonia), *reticulata* (Amazonia), *semifasciata* (La Plata, Mato Grosso), and *cametana* and *cyclostoma*, both from the Rio Tocantins/Xingu system. *Crenicichla sedentaria* was described by Kullander in 1986, and is apparently *'Batrachops'*-like in many respects. The name *sedentaria* refers to its sedentary habit of sitting on the bottom. All of the 'Hopping Pikes' are moderately to especially rheophilic (rapids-loving), have reduced swim bladders, and spend much of their time hugging the bottom and hopping, though all can swim. Those from the Tocantins/Xingu system prosper from the use of submersible powerheads.

Below: *This female* **Crenicichla sedentaria** *is a commonly-available 'Hopping' Pike from the* **Batrachops** *complex.*

THE CICHLASOMINES

Cichlasomines of the genus *'Cichlasoma'* have radiated dramatically in Central America from South American ancestors that crossed the isthmus of Panama several million years ago: there are now over 100 described species from Texas down through Panama. However, the number of Cichlasomines is decidedly more modest in South America proper, where the Acaras and their descendents have competed effectively for various ecological niches. Nevertheless, fish like the Angelfish, Discus and Uarau represent a few of the more creative evolutionary derivatives of the Cichlasomine lineage in South America.

Originally, fish having four or more hard spiny rays in their anal fins were relegated to the genus *Cichlasoma*. As was true for the three-spined Acaras, this was a rather heterogeneous collection of cichlids. It included, among others, a rather generalized, Acara-like fish named *Cichlasoma bimaculatum*, – the Guyanese 'Black Port' of the aquarium trade. Kullander realized that there were more similarities between the Port Acara *Aequidens portalegrensis* and the Black Acara than the simple difference in anal fin ray counts that divided them. In 1983, he lumped the two species and a number of other 'Port'-like species together in a common genus. The name of that genus was none other than *Cichlasoma*, since *C. bimaculatum* had been the author Swainson's original type-specimen for the genus in 1839!

This redefinition of the genus *Cichlasoma* made nomenclatural orphans of the rest of the Cichlasomine cichlids. For a while, the name *Heros*, next in historical line, was used in place of *Cichlasoma* for these orphaned cichlids. However, in 1986, Kullander redefined and restricted that name to the Severum-like Cichlasomines. In addition, he resurrected old names for many of the sub-groupings. These usages are reflected here. Unfortunately, many of the Cichlasomine cichlids, particularly the Mesoamerican species, have yet to be completely re-evaluated and, for the time being, should be referred to as *'Cichlasoma'* species, with quotation marks around the generic name. Note, however, that the aquarium literature has yet to follow suit, and that most of these fish can be accessed only with their older *Cichlasoma* nomen.

Cichlasoma portalegrense

Port Cichlid

● **Distribution:** La Plata Basin, Argentina, southern Brazil.
● **Length:** Can grow up to 15-20cm (6-8in).
● **Diet:** Omnivorous.
● **Sexing:** Essentially isomorphic. Ripe females are more rotund than males.
● **Aquarium maintenance and breeding:** One of the easiest of the biparental substrate-spawning cichlids, hardy, 'industrial-strength' fish whose maintenance and dietary requirements are quite basic and easily met. 'Ports' are exemplary parents, and may be maintained at lower temperature of 20-24°C (68-75°F) with brief drops to 15°C (60°F) because of their subtropical distribution. *Cichlasoma portalegrense* was historically one of the earliest cichlids in the hobby, making their appearance in Europe in the early 1900s.

Formerly known as *Aequidens portalegrensis*, under which name its care has been described for decades in the hobby literature, *Cichlasoma bimaculatum*, the Black Port, is a frequent import from Guyana, and several of the other *Cichlasoma* species *sensu stricto* (eg *C. taenia, amazonarum, paranense, araguaiense* and *dimerus*) are occasional accidental contaminants, and are usually sold as 'Ports'. Occasional 'movable

Above: *The Port Cichlid,* **Cichlasoma portalegrense,** *was one of the first cichlids to be exported from South America, in the early 1900s.*

platform' spawners, choosing leaves when available.

'Cichlasoma' festae

Red Terror

● **Distribution:** Pacific slope of Ecuador.
● **Length:** Males up to 46cm (18in), females grow to only two-thirds of that size.
● **Diet:** Omnivorous.
● **Sexing:** Conspicuously dimorphic. Males develop green iridescence on their flanks and in older specimens, 'craggy cheeks' (corrugated preopercle) and thick lips appear, eliciting the common native name for this particular species: 'vieja' or 'old woman'. Females of the species remain more rotund, smaller, and maintain the alternating bright red and black barring of juvenile fish.
● **Aquarium maintenance and breeding:** These fish have rightly earned their common name 'Red Terror': they are extremely belligerent. Adults must be housed in large tanks with rowdy cichlid tankmates. Even so, 'pairs' may routinely liquidate each other and the fish they are housed with. For larger specimens, the 'partial-divider' method is the best propagation strategy. Alternatively, a group of juveniles may be raised to sexual maturity and allowed to form

Below: *This courting female Red Terror,* **'Cichlasoma' festae,** *is absolutely incandescent.*

compatible pairs. They will spawn at 10-15cm (4-6in) despite their huge maximal size. These are typical biparental substrate spawners that lay several hundreds of eggs. Nothing compares with the incandescent coloration of a brood-tending female *'C.' festae* – the vertical bars turn bright orange-red alternating with black.

For all their natural aggression, these are gentle and excellent parents and a joy to watch in the aquarium. They should be handled like the majority of large Cichlasomines in terms of water maintenance and feeding.

The Cichlasomines from west of the Andes in northwestern South America are quite unlike those to the east, and more closely resemble their relatives from Central America. *'Cichlasoma' festae* clearly resembles the drabber Central American species, *'C.' uropthalmus*, both in coloration and habit and is often confused with it in the hobby, usually to the advantage of the seller. Two other generalized Cichlasomines from northwestern South America include *'C.' atromaculatum*, from the Rio Atrato in Colombia and the extremely rare *'C'. ornatum* (Colombia, Ecuador). At the moment neither of these fish is particularly common in the hobby.

Caquetaia spectabilis

Spectabile, False Basketmouth

● **Distribution:** Amazonia, Rio Tocantins/Rio Xingu system, Guyana.

● **Length:** Up to 25-30cm (10-12in) in the aquarium.

● **Diet:** Gape-and-suck ambush predators in the wild, omnivorous in the aquarium.

● **Sexing:** Essentially isomorphic.

● **Aquarium maintenance and breeding:** Beautiful but somewhat delicate fish. Particularly sensitive to water quality, they respond to lax maintenance by neuromast 'pitting'. Despite their apparent piscivory, they will prosper on a diet of pelleted prepared foods, freeze-dried krill and a variety of frozen foods. Earthworms may be useful in conditioning adults. Spawns are huge, numbering from 500-1000 depending on the size of the adults. Pairs of these fish are excellent parents to their fry.

Despite its widespread distribution, *C. spectabilis* is rather uncommon in the hobby and is rarely imported from the wild, which is unfortunate as it is one of the most beautiful of the South American Cichlasomines. There are two other species in the

Below: *The Spectabile,* ***Caquetaia spectabilis****, is a rare and somewhat delicate Cichlasomine.*

Above: *This female Chocolate Cichlid,* **Hypselecara temporalis.** *shown here tending eggs, lacks the nuchal hump of her mate.*

genus *Caquetaia. Caquetaia kraussii* hails from Venezuela and Colombia (Rio San Juan, Rio Atrato basins), and is a less colourful, bronze-brown version of *C. spectabilis.*

The third species, *C. myersi*, is also from the Colombian tributaries of the Amazon and Orinoco basins, but has been absent from the hobby. They share similar requirements in the aquarium, but only *C. spectabilis* and *C. kraussii* have been spawned and are occasionally available. The genus name *Caquetaia* derives from the Rio Caquetà in Colombia, the site of original capture of the type specimen for the genus. These fish were once placed in the genus *Petenia*, along with *P. splendida* from Central America, an ecological analogue with a similarly protrusible mouth and piscivorous habit.

Hypselecara temporalis

Chocolate Cichlid

● **Distribution:** Peruvian and Brazilian Amazon.
● **Length:** Up to 25cm (10in).
● **Diet:** Omnivorous.
● **Sexing:** Males with nuchal hood on forehead.
● **Aquarium maintenance and breeding:** Chocolate cichlids can be rather belligerent, particularly with their own. The difficulty in raising and spawning them, therefore, lies in establishing a compatible pair. While adults of both sexes can be paired by the 'blind date' method, a better solution is to raise a group of juveniles up to sexual maturity. Other, similarly aggressive, target fish help to cement the pair bond and dispel intra-pair aggression. In addition, this species is reproductively-precocious and will spawn at 10-13cm (4-5in), despite the much larger maximum adult size. Some attention should be paid to water quality and chemistry, but most foods are eaten ravenously and growth is rapid. Egg-eating is typical of young pairs but disappears with experience. Raising the fry is straightforward, and there is a ready market for the hundred of juveniles each spawning can generate.

Kullander created the genus *Hypselecara*, meaning 'High Acara', to commemorate the egg-shaped body of this fish. The chocolate cichlid has been known under several names both in science and in the hobby. These include *(Cichlasoma) goeldii, crassa, hellabrunni, coryphaenoides, arnoldi* and *Chuco axelrodi.* With the exception of *coryphaenoides*, now also in *Hypselecara*, the rest have been synonymized with either *H. temporalis* or *H. coryphaenoides.*

Typically, these have been geographic colorational variants elevated to species status. *H. coryphaenoides* is found in the Negro, Orinoco and Trombetas drainages and is readily distinguished from *H. temporalis.* The juvenile fish present a sharply-chiselled head profile, often with a light-coloured blaze down the ridge, and have a midlateral spot or band that extends vertically to the upper lateral line.

H. coryphaenoides is a species occasionally imported, but *H. temporalis* is the usual chocolate cichlid of the hobby, being bred commercially in Florida and Asia.

Heros severus

Severum

● **Distribution:** Guyana, Amazonia, Colombia, Venezuela.
● **Length:** Grows up to 25-30cm (10-12in).
● **Diet:** Omnivore, vegetable material recommended.
● **Sexing:** Essentially isomorphic. Males more elongate than females. In some populations, males with more extensive vermiform spotting on flanks.
● **Aquarium maintenance and breeding:** Care as for the chocolate cichlid, *Hypselecara temporalis.* Water should be kept clean and warm around 26-29°C (78-84°F), and the fish fed a variety of plant material (eg romaine, lettuce, spinach) in addition to the regular prepared pelleted, frozen and live foods (earthworms). Severum can be belligerent, especially as adults. Compatible pairing may be difficult with the 'blind date' method, even if the adults are kept divided for some time.

The presence of target fish may increase the chances of success. These are typical biparental substrate spawners which spawn precociously at a size of 10cm (4in) at between 12-18 months of age. There is always a good market in the hobby for young of this species, particularly the cultivar gold (xanthistic) variety.

Kullander, in 1986, restricted usage of the genus name *Heros* to the species *severus* and elevated at least one colorational variant from Peru, to species status (eg *H. appendiculatus*). It is likely that other geographic variants will receive species status in the near future.

In addition to wild specimens, *H. severus*, both as green and xanthistic gold forms, are bred in Asia and Florida and are freely available to the aquarist.

Below: *The Green Severum,* **Heros severus**, *is commonly available in the trade.*

Above: *The Flag Cichlid,* **Mesonauta festivus,** *is found with Angelfish in the wild.*

Mesonauta festivus

Festivum, Flag Cichlid

● **Distribution:** Guianas, Orinoco and Amazon basins.
● **Length:** Grows up to 15-20cm (6-8in).
● **Diet:** Omnivorous.
● **Sexing:** Essentially isomorphic. Males with longer 'nose'.
● **Aquarium maintenance and breeding:** Festivum are typically found in association with Angelfish, *Pterophyllum scalare*, in the wild, and should be maintained like wild angelfish. Soft, acid water is helpful, but not essential. Because of their shyness, they should be kept in a planted tank with some shelter, perhaps bogwood. They neither dig nor eat plants. They are relatively peaceful and can be kept with other less-aggressive cichlids, like Eartheaters or Acaras, or dithers large enough to escape being eaten. Spawns are of modest size (300-500 eggs) and pairs are reliable parents once induced to spawn, which can be difficult. Males often have a larger, more exaggerated 'Roman nose'.

A nice alternative to the larger and more aggressive South American Cichlasomines.

Kullander, in 1986, resurrected the genus *Mesonauta* for the species *festivus*. Noting the extreme geographic variability of Festivum, Kullander, in 1991, split the species into five. These include *insignis* from the upper Rio Negro and Rio Orinoco, *acora* in the Tocantins and Xingu drainages, *egregius* in the Columbian Orinoco basin, *mirificus* in the Peruvian Amazon, and *festivus,* now restricted to the Paraguay and Bolivian Amazon basins, the Rio Jamari and lower Rio Tapajos. They differ in body proportions and markings, and there are several subtle differences in spine count, etc.

At the time of this publication, Guyanan and Brazilian Amazonian specimens have not yet been studied, but it is likely that Kullander will eventually treat these as discrete species.

Most of the commercially available Festivum are bred in captivity in Asia or Florida, with the remainder available as wild fish, usually originating from Guyana.

Hoplarchus psitticus

Parrot Cichlid

● **Distribution:** Amazon and Orinoco basins.

● **Length:** Older adults to at least 40-46cm (16-18in).

● **Diet:** Omnivorous, some vegetable material is recommended.

● **Sexing:** Essentially isomorphic. Males slightly more elongate than females with a slight nuchal crest as they grow large.

● **Aquarium maintenance and breeding:** Care as for the generality of blackwater species. Soft, acid water kept exceptionally clean and warm, 26-30°C (78-86°F), is a must: these fish are particularly susceptible to neuromast erosion (Head Hole). Adults can be aggressive towards each other and pairing can be a problem.

They have been spawned only a few times in the aquarium. Eggs number from 200-300 and are placed on a vertical surface. Because the parents are prone to eating their spawn, what fry have been raised have been done so artificially. The fish are quite shy and easily distracted from spawning by the presence of other fish, or by placement of their tank in a public venue. It seems also that reproductive maturity is reached only after considerable age at about 3-5 years.

Available only very infrequently, typically as accidental contaminants with chocolate cichlids, *Hoplarchus psitticus* is apparently rare in the wild. The Amazonian specimens are the more colourful thant their Colombian counterparts. Inclusion of carotene-containing foods (eg krill) will help bring out the bright red highlights on this otherwise iridescent green fish. They also appreciate some 'greens' in their diet.

'Cichlasoma' facetum

Chanchito, Zebra Cichlid

● **Distribution:** Southeastern Brazil, Argentina, Uruguay, La Plata drainage.

● **Length:** Grows up to 15-20cm (6-8in).

● **Diet:** Omnivorous.

● **Sexing:** Essentially isomorphic.

● **Aquarium maintenance and breeding:** Despite its smaller size, the Chanchito exhibits all of the behavioural quirks of the larger Chiclasomines: it is aggressive and it digs. However, it makes a reasonable addition to a not overly rowdy community of cichlids. Chanchito means 'little pig' and aptly summarizes the feeding behaviour of this fish: they are aggressively and enthusiastically

Below: *The Parrot Cichlid,* **Hoplarchus psitticus,** *grows to around 40cm (16in).*

Above: *Despite its smaller size, the Chanchito,* **'Cichlasoma' facetum,** *can be as aggressive as its larger relatives.*

omnivorous. Breeding is as for the generality of biparental substrate-spawning cichlids. They are easily induced to spawn and are generally excellent parents. Hailing from subtropical South America, these fish can take lower temperatures of 20-24°C (68-75°F), even briefly down to around 13°C (55°F). This is one reason why the Chanchito was historically the first successfully kept aquarium cichlids in the mid-1890s.

There are more 'Chanchito-like' fish which may well be distinct species: *'Cichlasoma' autochthon* and *'C.' oblongum,* somewhat smaller than *'C.' facetum* at 13-15cm (5-6in) and with vertical barring of variable colour and intensity. This fish is not propogated commercially, consequently we must rely on infrequent accidental importations from Argentina.

Pterophyllum altum

Altum Angelfish

● **Distribution:** Rio Orinoco, Rio Negro.
● **Length:** Up to 20-25cm (8-10in).
● **Diet:** Omnivorous. Frozen and live foods recommended.
● **Sexing:** *Pt. altum* are essentially isomorphic.
● **Aquarium maintenance and breeding:** *Pterophyllum altum,* the deep-bodied relative of the more common angelfish, *Pterophyllum scalare*, is occasionally imported from Colombia and may be distinguished from *Pt. scalare* by its higher body, peculiar long, curved snout, and somewhat indistinct brown vertical barring. Angelfish, in general, are found both in whitewater or turbid blackwater, usually in lakes or other lentic habitats, typically with densely vegetated shores punctuated with fallen submerged branches. Whereas *Pt. scalare* yielded to the breeder's skill in the early 1900s and is today the staple of the aquarium hobby, having been captively inbred with strains selected and 'fixed' for coloration and finnage mutations over the past 80 years, *Pt. altum* has to date resisted all attempts. For the serious hobbyist wishing a challenge, the Altum Angelfish is it! Generous feedings of live foods and soft, RO-processed blackwater kept warm, and very clean water will undoubtedly be required.

Pterophyllum altum are best maintained as a colony in their own separate, large tank, chosen to maximize front-to-back depth and with adequate bogwood shelter and planting.

Below: *The Altum Angelfish,* **Pterophyllum altum,** *is a delicate though beautiful species.*

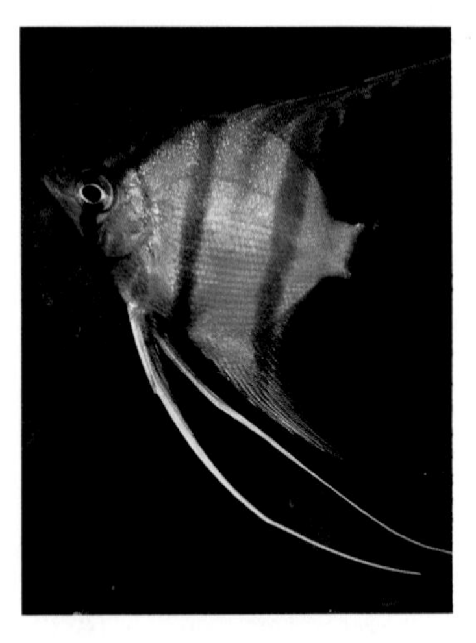

Left: *The common Scalare Angelfish,* **Pterophyllum scalare,** *lacks the curved snout of its congener.*

There is one other recognized species of Angelfish: *Pt. dumerilii* from the Amazon. Kullander (1986) believes *dumerilii* to be a junior synonym of *scalare* and recognizes the species *Pt. leopoldi* as valid. It is conspicuous for the dark black spot located at the dorsal tip of the fourth vertical bar, just near the site of dorsal fin insertion. It is also rather 'dumpy' (low-bodied), relative to both other species.

Symphysodon aequifasciatus

Discus, Pompadour

● **Distribution:** Blackwater Amazonia, Tocantins/Xingu system.
● **Length:** Up to 20-25cm (8-10in).
● **Diet:** Frozen and live foods.
● **Sexing:** Essentially isomorphic. In some populations (eg blue), males with more pronounced vermiform markings on flanks.
● **Aquarium maintenance and breeding:** Like the Angelfish, *Pterphyllum scalare*, Discus have become readily available in the hobby as tank-raised cultivars whose requirements are best learned from speciality books. Modern Discus are very different fish than their wild progenitors. For one, they are much more easily induced to propagate in the aquarium and they tolerate a much wider range of water chemistries and conditions. Moreover, they are usually kept in the sparest of conditions, typically, bare tanks with a slate or spawning cone to hold their eggs. Wild Discus should be treated as the blackwater species they are and kept in soft, acid, warm water, around 27-30°C (80-86°F), with added peat extract and kept very clean with effective biological filtration and regular water changes. They prefer tanks with larger bottom areas and shelter in the form of a tangle of bogwood and/or plants to offset their shyness. Dither fish, principally tetras, make them feel more secure. Peaceful cichlids, including *Apistogramma* species, may also be kept with them. One tankmate that should be excluded is wild Angelfish of any species which often carry diseases fatal to Discus.

Discus prosper on a diet of frozen insect larvae (bloodworms, mosquito, glassworms) and live foods. Spawning is another matter, however. Wild Discus are difficult to induce to spawn in the aquarium. A rich diet and RO-processed water kept warm and clean are helpful. Nevertheless, spawnings of wild fish still occur, and many of the various coloured cultivated strains have been engineered by occasional and calculated outcrossing with selected exceptional wild fish. When they do spawn, wild pairs often make exceptional parents and support the initial growth of their fry via contact feeding – a behaviour pattern in which the young fish 'nip-off' nutritional mucus from both parents' sides. Growth, thereafter, is rapid on a diet of newly-hatched *Artemia* nauplii.

The several colour varieties of *S. aequifasciatus,* brown, blue and green, were given subspecific status by several

Above: *The Discus,* **Symphysodon aequifasciatus,** *is the king of aquarium fish.*

ichthyologists (eg *a. axelrodi, a. haraldi, a. aequifasciatus*), but the modern view is that they are one species with several geographic variants, all of which can interbreed. A second species, *S. discus,* better known as the True or Heckel Discus, seems distinct. The Heckel Discus, found also in the Amazon, is characterized by somewhat larger and higher body, flanks completely vermiculated by iridescent stripes, and three distinct black vertical bars, the thickest and most distinct through the centre of the fish. These also come in several colour varieties. *Symphysodon aequifasciatus* and *S. discus* have occasionally been cross-bred and the resultant offspring have proven fertile, calling into question the biological distinctiveness of the two species. However, wild hybrids have not been found. Care of *S. discus* is identical to that of *S. aequifasciatus*; however, they seem to enjoy higher temperatures of up to 32°C (90°F).

Uaru amphiacanthoides

Uaru, Triangle Cichlid

● **Distribution:** Amazon drainage.
● **Length:** Can grow to at least 30cm (12in).
● **Diet:** Omnivorous, vegetable material recommended.
● **Sexing:** Essentially isomorphic.
● **Aquarium maintenance and breeding:** Often associated with Discus that form colonies around submerged trees or branches. Care is that of Discus. They require warm 26-30°C (78-86°F), very clean, soft, acid water. Because these are rather dramatic herbivores, vegetable matter in the form of lettuce or spinach should be provided regularly. Uaru are biparental substrate spawners and have been spawned regularly in the aquarium, but are notorious egg eaters.

Often, two females will 'pair up' and spawn regularly, alternating egg-laying roles. Artificial incubation of eggs may be necessary to prove the identity and fertility of the 'pair'. When and if they get it right, Uaru are excellent parents, providing the first nutrition for their brood in the form of nutritional mucus that the fry

Above: *The Triangle Cichlid,* **Uaru amphiacanthoides,** *requires lots of vegetable material in its diet.*

graze from the surface of the flanks: Uaru, like Discus, contact feed. Although sometimes called 'Poor Man's Discus', Uaru are highly desirable and sought-after cichlids in their own right, and there is a steady market for tank-reared young. Pairs are best had by raising a group of juveniles to sexual maturity. One interesting aspect of their development is the colorational metamorphosis that youngsters undergo as they mature. Juvenile Uaru are dark brown with regularly spaced white blotches scattered on their flanks. As they mature, these spots enlarge to blotches as the base colour lightens to café-au-lait brown. The adult fish retain this lighter brown colour and develop a black 'eyebrow' mark along with the large black flank triangle.

Below: *Juvenile Uaru go through a distinctive colorational metamorphosis as they mature.*

Uaru species have been reported from the Tocantins/Xingu system and from Venezuela. The latter was described as *U. fernandezyepezi* by Stawikowski in 1989. One undescribed species has a rectangular blotch replacing the triangle on the flanks, and lacks the black 'eyebrow' marking.

MISCELLANEOUS CICHLIDS
There are several South American cichlids which defy easy placement within any of the major assemblages. These include the oscar, various basketmouth cichlids, and the predatory peacock bass.

Cichla sp. affin. ocellaris

Lukanani, tucanaré, Peacock Bass

● **Distribution:** Amazon basin, Orinoco basin, Tocantins/Xingu system, Guianas.
● **Length:** Up to 61cm (24in) in the wild.
● **Diet:** Piscivore. Will convert to krill and frozen fish.
● **Sexing:** Essentially this species is isomorphic. Large 'bull' males reportedly develop a pronounced nuchal hump just prior to spawning.
● **Aquarium maintenance and breeding:** Because of their large size, *Cichla* species have been kept only by the most devoted of specialists. As 'cute' juveniles they seem to accept only live fish or live Tubificid worms. Adults may, with some trouble, be moved on to a diet of freeze-dried krill, and frozen prawns. They are messy eaters and the water must be kept clean and moving. Reports by Lowe-McConnell (1969) suggest that in the wild *Cichla* species pairs dig several circular nests on the bottom and lay between 6000 and 10000 eggs in one of them. They are devoted parents and guard their young until they reach a length of 4cm (1½in).

The large ocellated tail spot provides an orienting signal for the fry. Additionally, this eye-like spot is believed to confuse piranhas who tear pieces of fins from fish. These cichlids are recommended only for aquarists with the largest of tanks and much dedication. Perhaps this species is best left in South America where they belong.

Historically, two species of *Cichla* were recognized to science: *C. ocellaris* and *C. temensis*. Recently, Kullander (1986) has proposed splitting *ocellaris*, which exhibits truly dramatic regional polymorphism, into a series of discrete species including

Below: *The tucanaré,* **Cichla sp. affin. ocellaris,** *is a prized game fish throughout South America.*

monoculus (Peru, Amazon), *intermidia, orinocensis* (Venezuela) and, of course, *ocellaris* (Guianas). In the absence of reliable information as to the orgin of export, the name *Cichla sp. affin. ocellaris* is best applied to this fish. *Cichla* species are beloved food and game fish in their native South America, and have been introduced into Central America for the same purposes.

Astronotus ocellatus

Oscar

● **Distribution:** Amazon and Orinoco basins, French Guiana, Northern Paraquay. Some of these may be distinct species other than *ocellatus*.

● **Length:** Up to 30-36cm (12-14in).

● **Diet:** Omnivorous.

● **Sexing:** Essentially isomorphic.

● **Aquarium maintenance and breeding:** Unfortunately, young Oscars are often among the first fish new hobbyists acquire for their 45-litre (10-gallon) community tank

Above: **Cichla temensis** *is notable for its body spotting.*

because, at 2.5-5cm (1-2in), their large eyes and social demeanour make them 'so cute': too bad for the Oscars because they are potentially large fish with special requirements. Usually they perish, and the ornamental trade and breeders count on that eventually to ensure repeated replacement. If treated well, an Oscar can live for ten years or more, and will outgrow its 10-gallon tank in a few months. Young Oscars should be raised as a group (it is a gamble to attempt to 'blind date' large individuals) in as large a tank as possible. The water should be heavily filtered, as these are messy eaters, and partial regular water changes must be done. Lax attention to water quality inevitably results in neuromast erosion (pitting) which is usually not reversible. The bigger the tank, the less of a potential problem this is. Oscars will readily take pelleted prepared foods and freeze-dried krill and do not need living feeder fish to prosper. The Oscar is prone to rearranging its tank, so care should be taken in the placement of rock piles that could be undermined, fall and ultimately damage the fish or crack the tank. Heaters should be placed and fastened with an eye to destruction or jettisoning. As they grow and reach reproductive maturity in 1-1½ years, they begin pairing off. Housing a pair or group in a 570-litre (125-gallon) tank with compatible target cichlids is not overkill. Once bonded, they may stay together for as long as ten years, spawning all the while. They lay 1000 or so opaque, white eggs (good eggs look like bad, fungused eggs) on a previously cleaned substrate and both parents participate in their care and defence. The fry grow rapidly and are eminently saleable at a size of 2.5cm (1in) for the reasons outlined above. The Oscar is a 'personality' fish and soon learns to interact with its owner.

Below: *The Oscar,* **Astronotus ocellatus,** *is a large 'personality' fish which endears itself to its owners.*

Kullander (1986) suggests that there are several distinct species of *Astronotus* as yet to be described as well as a second recognized species: *A. crassipinnis*, from Peru. In general, these differ in fin and ray counts, and in colorational pattern (ie absence, presence and extent of orange ocellations). The Oscar is another cultivated ornamental fish which has been selected for mutations in coloration and finnage. The Tiger and Red Oscars are two such examples where the extent of red coloration has been improved on over wild fish. More recently, an albino strain has been developed and fixed. In addition, a 'veil-tailed' mutation has been introduced into most of the colour types. These are typically bred in Asia and Florida and are the primary source of hobby Oscars. These latter are particularly tolerant of a wide range of

water chemistries. Wild specimens, which are considerably less colourful, are only occasionally imported, and are considerably less forgiving.

Acaronia nassa

Basketmouth

● **Distribution:** Guianas, Amazon Basin.
● **Length:** Up to 20cm (8in).
● **Diet:** Piscivorous, can be shifted to freeze-dried krill or frozen foods.
● **Sexing:** Essentially isomorphic; ripe females heavier, less elongate than males.
● **Aquarium maintenance and breeding:** The Basketmouth makes its living as a gape-and-suck ambush piscivore in the wild, hence its common name. The protrusible mouth can be extended in a flash, and the vacuum created is sufficient to suck in swimming prey quickly and decisively. In the aquarium, Basketmouths will learn to take large freeze-dried krill, earthworms and even some pelleted foods, but living feeder-fish may be useful for conditioning potential breeders. They are undemanding about water chemistry, but attention should be paid to water quality. Adults may be somewhat aggressive with each other, but the Basketmouth can be successfully kept in mixed cichlid communities. They do require shelter into which they can retreat. Captive spawning has been reported only once (Leibel, 1985), and in that instance the fish behaved as a typical biparental substrate spawner. Spawning was effected by cyclical temperature increases to 33°C (92°F) and decreases to 22°C (72°F) every few days over the course of several weeks. The fry grew quickly on an initial diet of newly-hatched *Artemia*, and were switched to diced frozen bloodworms within two weeks. Basketmouths are rarely seen in the hobby, entering typically as single.

A second species, *Acaronia vultuosa*, has recently been described from the Orinoco drainage, which differs principally in the spotting pattern of the head.

Chaetobranchus flavescens

Combtail Basketmouth

Below: The Basketmouth, **Acaronia nassa**, *is a gape-and-suck ambush predator.*

Above: *The Combtail Basketmouth,* **Chaetobranchus flavescens**, *has never been bred in captivity.*

● **Distribution:** Amazon drainage.
● **Length:** Up to 25-30cm (10-12in).
● **Diet:** Freeze-dried krill and a variety of frozen and prepared foods.
● **Sexing:** Essentially isomorphic. Both sexes have dramatically produced fin filaments on the dorsal and caudal fins. Males slightly more elongate.
● **Aquarium maintenance and breeding:** Spawning habits unknown as it has yet to be captively propagated. Like the 'true' Basketmouth, these have trapdoor mouths; however, the generic name refers to their long, thin gill rakers which suggest a life of plankton-sifting for this fish. Luckily, *Ch. flavescens* readily take freeze-dried krill and a variety of offered frozen foods in the aquarium. Care as for the *Acaronia nassa*. Adults of both sexes develop long filamentous streamers on the edges of their tail fins which are echoed in the multi-filamented dorsal streamers, hence the common name.

A second species in the genus, *Ch. semifasciatus*, is occasionally imported all the way from the Brazilian Amazon and may be distinguished from *Ch. flavescens* by the presence of a prominent ocellus on the caudal preduncle and several partial vertical bars on the flanks. A second closely related genus, *Chaetobranchopsis*, with three valid species – *australis*, (Argentina, Paraquay), *bitaeniata* (Amazon) and *orbicularis* (Amazon) – is distinguished on the basis of fin ray counts, but, like *Chaetobranchus*, members have the setiform gill rakers characteristic of plankton feeders. These fish apparently do not make their living straining small zooplankton and, thus, do poorly in the aquarium.

Occasionally, exceptional individuals can be converted to powdered krill or even larger types of food, but typically, *Chaetobranchopsis* species in captivity suffer a long, protracted starvation death.

Index

Page numbers in **bold** indicate major references, including accompanying photographs. Page numbers in *italics* indicate captions to other illustrations. Less important text entries are shown in normal type.

D

Q

R

S

T

Bibliography and Further Reading

Cichocki, F., *Cladistic History of Cichlid Fishes and Reproductive Strategies of the American Genera Acarichthys Biotodoma* and *Geophagus* Dissertation, University of Michigan (USA), 1976
Fryer, G. and Iles, T.D. *The Cichlid Fishes of the Great Lakes of Africa* Oliver and Boyd, Edinburgh, 1972
Gosse, J.P., *Revision du Genre Geophagus (Pisces: Cichlidae) Memoires Académie Royale Science d'Outre Mer (Brussels)* **19(3)**: *1-172, 1975*
Goulding, M., *Amazon – The Flooded Forest*, BBC Books, London 1989
Knoppel, H.A., *Food of Central Amazonian Fishes. Contribution to the Nutrient Ecology of Amazonian Rain-Forest Streams Amazonia* **2**: 257-352, 1970
Kullander, S.O., *A Revision of the South American Cichlid Genus Cichlasoma (Teleostei: Cichlidae)* Monograph, Swedish Museum of Natural History, Stockholm, 1983.
Kullander, S.O., *Cichlid Fishes of the Amazon River Drainage of Peru* Monograph, Swedish Museum of Natural History, Stockholm, 1986
Kullander, S.O. and H. Nijssen, *The Cichlids of Surinam (Teleostei: Labroidei)* E.J. Brill., The Netherlands, 1989
Leibel, W.S., *Movable Platform Spawning in an Increasing Number of Neotropical Cichlids* Buntbarsche Bulletin, Journal of the American Cichlid Association, **107**: 2-10, 1985
Leibel, W.S., *Behavioral Evidence for the Polytypic Nature of Geophagus surinamensis (Bloch 1791)* Buntbarsche Bulletin, Journal of the American Cichlid Association, **113**: 2-15, 1985
Leibel, W.S., *Tips for the Maintenance of Neotropical Geoghagine Cichlids* Buntbarsche Bulletin, Journal of the American Cichlid Association, **109**: 3-7, 1985
Lewis, D., Reinthal, P., Trendall, J., *A Guide to the Fishes of Lake Malawi National Park* World Wildlife Fund
Linke, H. and Staeck, W. *Afrikanische Cichliden. I. Buntbarsche aus Westafrika* Tetra-Verlag, Melle, 1981
Loiselle, P.V. *The Cichlid Aquarium* Tetra Press, Melle, 1987

Lowe-McConnell, R., *The Cichlid Fishes of Guyana South America with Notes on their Ecology and Breeding Behavior,* Journal of the Linnaean Society (Zoology) **48**: 255-302, 1969
Ribbink, A.J. et al. *A preliminary survey of the cichlid fishes of rocky habitats in Lake Malawi* S. Afr. J. Zool. **18(3)**: 149-310, 1983
Stawikowski, R. and Werner, U., *Die Buntbarsche der Neuen Welt: Sudamerika* Reimar Hobbing Verlag, Germany, 1988

Specialist Societies

American Cichlid Association
Membership Committee
PO Box 31230
Raleigh, NC 27622
USA

British Cichlid Association
Membership Secretary
16 Kingsley Road
Bristol BS6 6AF
UK

Picture credits

Photographs
The publishers wish to thank the following photographers and agencies who have supplied photographs for this book. The photographs have been credited by page number and position on the page: (T) Top, (B) Bottom, (C) Centre.

David Allison: 73(T), 107, 115, 137

Dr Chris Andrews: 73(B)

Eric Crichton © Salamander Books Ltd: 66

Lee Finley: 229

Interpet Ltd: 76, 77

Jan Eric Larsson: 22-23, 42-43, 62-63, 68, 88-89, 102-103, 104, 105, 109, 110, 111, 114(B), 120, 124, 126, 129, 135, 138, 141, 144, 148

Dr Wayne S Leibel: 18, 19, 29, 71, 94-95, 96, 192(T), 193, 195(T), 196(B), 202, 206, 207, 208, 214, 220, 221, 223, 224, 226, 227, 228(B), 231, 233

Dr Paul V Loiselle: 12, 41, 97, 99, 113, 117(B), 118, 125, 132, 136, 139(T,B), 140, 142, 143(T,B), 145, 146, 147, 149(T), 150(T,B), 192(B), 203, 204(T), 211(T), 219(B)

Arend van den Nieuwenhuizen: 83, 86, 106, 114(T), 123, 130-131, 151, 156, 172-173(B)

John O'Malley: 59, 70, 79, 98, 188-189, 190, 194, 195(B), 196(T), 197(T,B), 199, 200, 201, 204(B), 205, 210, 211(B), 213(T), 215, 216, 217(T,B), 219(T), 222, 225(T,B), 228(T), 230, 232

Laurence Perkins: 157

Salamander Books: 64

Mike Sandford: 117(T), 128, 168, 185(B)

David Sands: Half title, Title, 4-5, 6-7, 8-9, 10-11, 24, 44, 51, 60-61, 67, 89(B), 90(B), 91, 100-101, 155(T), 158-159(T), 162, 163(T,B), 167, 169(B), 174, 179(T), 185(T)

Ian Sellick: 152-153

William Tomey: 47, 72, 121, 122, 127, 149(B)

Uwe Werner: 25, 26-27, 30-31, 50, 57, 90(T), 92, 154, 155(B), 159(B), 160-161, 164-165, 166, 169(T), 170-171, 173(T,C), 175(T,B), 177, 178-179(B), 180(T,B), 181, 182(T,B), 183, 184, 186, 187(T,B)

Kurt Zadnik: 198, 198, 212-213(B)

Rudolf Zukal: 176